VISIONS OF HEAVEN

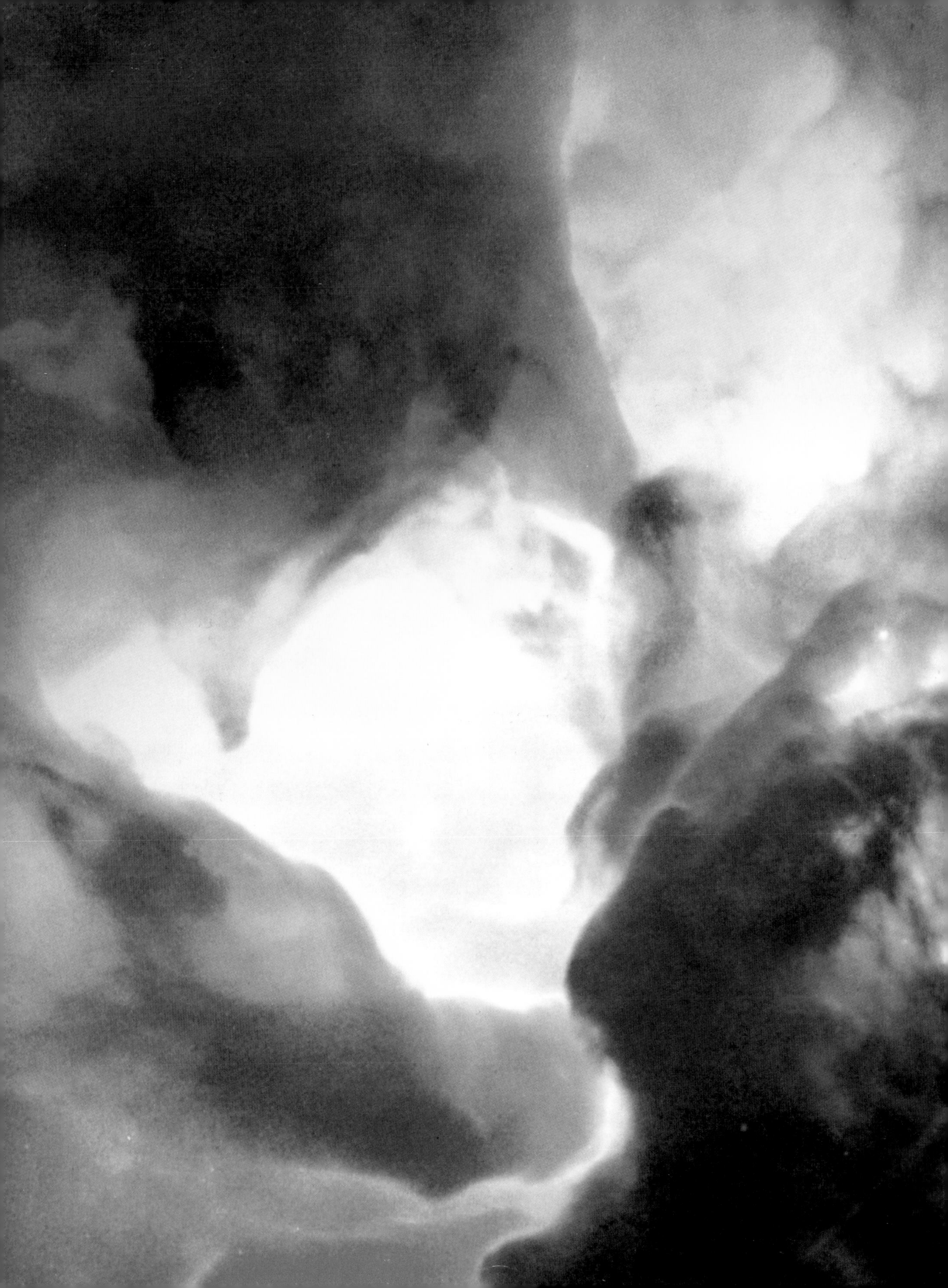

VISIONS OF HEAVEN

THE MYSTERIES OF THE UNIVERSE REVEALED BY THE HUBBLE SPACE TELESCOPE

Tom Wilkie and Mark Rosselli

Hodder & Stoughton

To my father. M.R.
To my wife. T.W.

FRONTISPIECE: A close-up of the Lagoon nebula showing the giant 'twister' created by the action of Herschel 36. *Credit: A. Caulet (ST-ECF, ESA), and NASA*
PAGE 6: The 'hour-glass' nebula, MyCn18. *Credit: R. Sahai, J. Trauger (JPL), the WFPC2 science team, and NASA*
PAGE 8: A close-up of a region of the Eagle nebula where young stars are being incubated. *Credit: Jeff Hester and Paul Scowen (Arizona State University), and NASA*

1 3 5 7 9 10 8 6 4 2

BRITISH LIBRARY CATALOGUING IN PUBLICATION DATA

A CIP catalogue record for this title is available from the British Library.

ISBN 0 340 71734 3

Printed and bound in Great Britain by
Butler & Tanner Ltd, Frome and London

HODDER & STOUGHTON
A division of Hodder Headline PLC
338 Euston Road
London NW1 3BH

CONTENTS

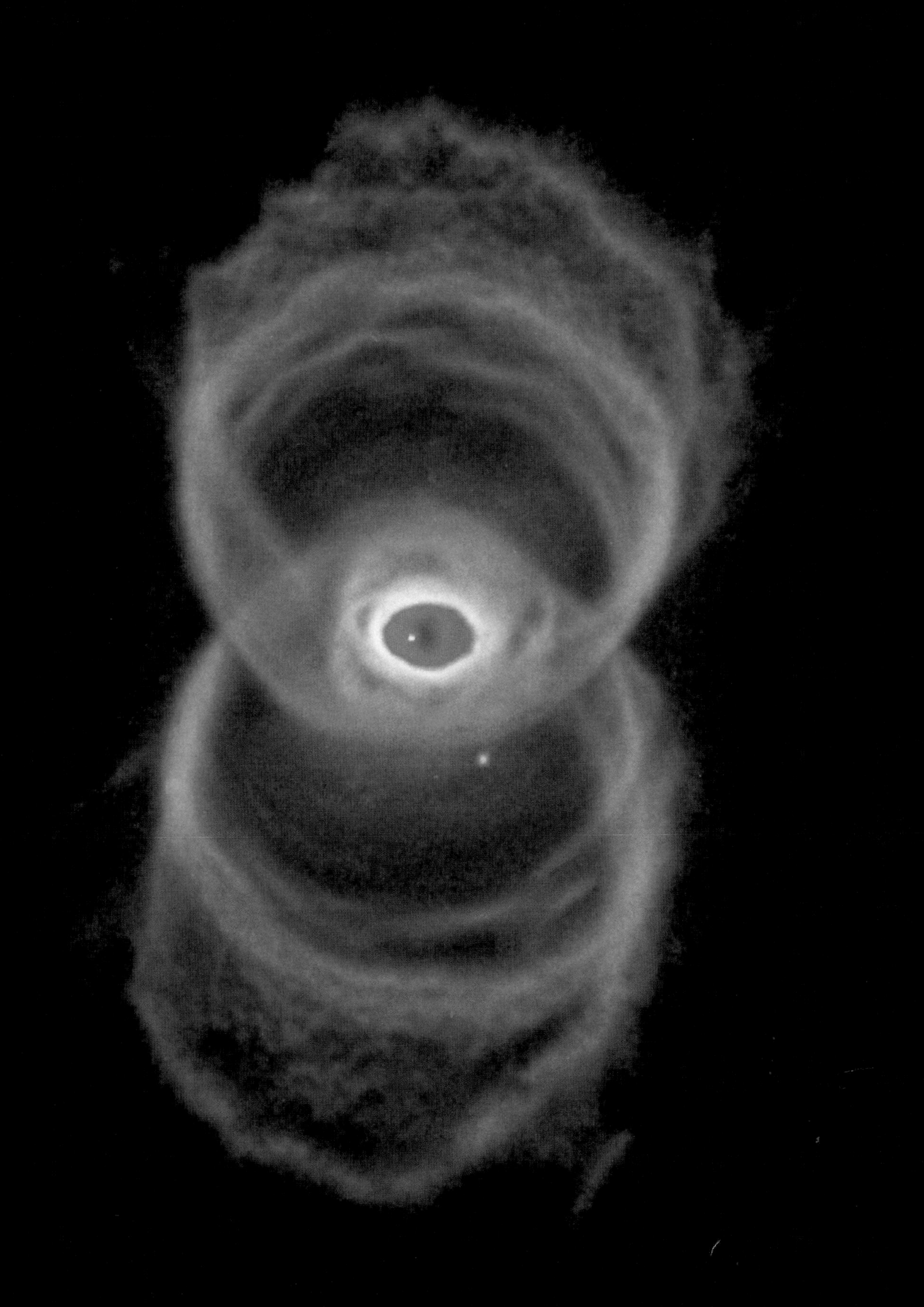

FOREWORD

by The Astronomer Royal, Sir Martin Rees

An undoubted scientific highlight of the 1990s has been the progress in mapping the cosmos. Astronomers are beginning to understand how our Sun and solar system formed, how they relate to the vast concourse of other stars of the Milky Way and how this entire system is just one galaxy among billions of others. Issues that were once entirely speculative have now come within the scope of serious science: we can, for instance, trace cosmic history, at least in outline, back to the initial instants of a so-called Big Bang that set our entire universe expanding.

This progress owes little to armchair theorists: it results from extensive observations using advanced technologies. Among the new instruments, the Hubble Space Telescope undoubtedly takes first place, as a technical achievement, and in its impact on our knowledge. Circling the Earth, high above the blurring and absorption of the atmosphere, it has yielded sharper images of our cosmic environment than we had before, clarifying our view of planets, stars and nebulae. Its ultra-sensitive detectors reveal galaxies so distant that their light has taken more than 10 billion years on its journey towards us. It thus gives us direct 'snapshots' of what our universe was like when it was young.

The Hubble Space Telescope has been crucial in helping astronomers to understand our cosmic origins. But its marvellous pictures have impacted on the consciousness of a wider public, just as the first views of our Earth, taken by the Apollo astronauts, did thirty years ago.

Many of these pictures are reproduced here. Tom Wilkie and Mark Rosselli offer a vivid commentary on the objects portrayed, setting them in context. The text offers, as well, a fascinating history of the decades of international effort that the Hubble Space Telescope involved – its conception, the vicissitudes and delays and the eventual scientific triumphs.

It is in the nature of science that every achievement leads to new questions and poses new challenges. This book is an eloquent celebration of a great chapter in science and leaves the reader in eager anticipation of the further progress that can be expected in the next millennium.

Sir Martin Rees is Astronomer Royal and Royal Society Research Professor at the University of Cambridge.

INTRODUCTION

Every picture tells a story and one picture is worth a thousand words. In this book, we have used the pictures taken by the Hubble Space Telescope to tell the story of the universe in which we live. Come with us on an unhurried tour of the cosmos. Pause and look at the sights, for many of them are breathtaking in their beauty.

This is neither a textbook force-feeding facts, nor a collection of pretty pictures. We have tried to combine Hubble's pictures and our words to tell a coherent story. And there are many stories to be told: of the birth and the death of stars; of how planetary systems might have come into being; of the great cities of stars we know as galaxies; and of the beginning and the end of all things – the universe itself.

Astronomers have long since abandoned taking pictures using photographic film. Hubble, like all modern telescopes, registers images electronically and beams back to Earth a stream of binary information which can then be reassembled to produce the visual images printed here. Some of these images are 'false colour'. Rather as dogs can hear a whistle whose pitch is too high for human ears, Hubble can 'see' colours which are invisible to us: lower than the red and higher than the violet ends of the spectrum. Such Hubble images have been transposed in wavelength – visual pitch, so to speak – so that we can see them.

The electronic data gives astronomers immediate access to numerical information about the intensity of light and its wavelength. By feeding the numbers into equations modelled on computer, they can check the accuracy of their theories about the universe. Although this book has not gone into the mathematics, it is the detailed numerical agreement that proves the theories are right. For the astronomers, seeing is no longer believing.

As will be evident from many places in the text, astronomers use data from radiotelescopes, infra-red telescopes and satellites, ultraviolet, X-ray, and gamma-ray observations, as well as from visible-light telescopes to expand their picture of the universe. Although we have used Hubble photographs to tell our story, Hubble by itself would not have yielded all we now know about our universe. The focus on Hubble pictures in this book should not mislead: these other visions are as vital. But perhaps the invisible cosmos is for a future writing.

CHAPTER ONE

IN THE BEGINNING

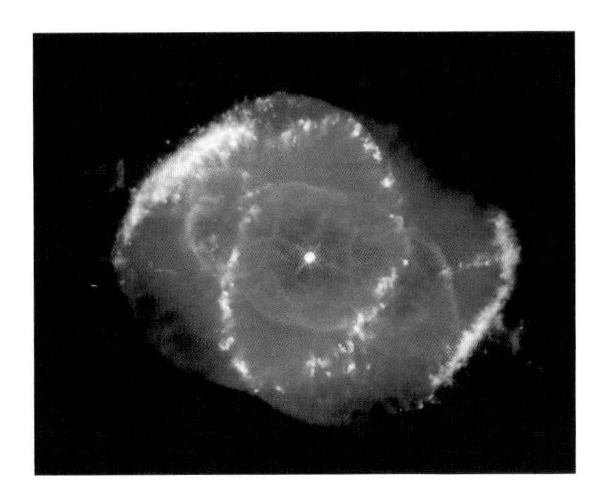

Each evening, as the Sun's radiance slowly fades from the sky, men and women across the world stop what they are doing and gaze up at the emerging stars. There is no reason to suppose that humans have ever done otherwise, from the moment we first stopped paying attention to objects solely because we wanted to eat them or because we feared they might want to eat us. And now, as then, the stars shine down a cool, clear light . . . silent, unchanging.

Opposite: This vast panorama of the heart of the Orion nebula was created by piecing together a mosaic of 15 separate fields. The entire image – which covers an area of the sky equivalent to 5 per cent of the area taken up by the full moon – measures 2.5 light years across. It shows a star factory in full production, with more than 700 young and baby stars in various stages of birth and early life. The nebula is located in the sword region of the constellation of Orion the Hunter.

Credit: C. R. O'Dell (Rice University), and NASA

Above: The 'Cat's Eye Nebula' shows part of the death throes of a star as it blows off some of its outer layers into space to form what is misleadingly called a planetary nebula. (It has nothing to do with the formation of planets.) In fact, astronomers think that this might be a double-star system because the structures revealed by the Hubble Space Telescope (HST) are much more complicated than those seen in most planetary nebulae. Not even the HST can separate the two stars which appear as a single point of light at the centre. The nebula is 3,000 light-years away in the northern constellation Draco, the Dragon.

Credit: J.P. Harrington and K.J. Borkowski (University of Maryland), and NASA

Five thousand years ago Egyptian priests, clustered on the roofs of their temples at Memphis, ordered their calendar by the cycle of the constellations and knew that the unfailing appearance of Sirius, the dog star, signalled the most important event in Egyptian agriculture: the flooding of the Nile. The gods of Ancient Egypt are long since dead; their temples are tumbled to dusty ruin; the mighty Nile itself has been tamed and no longer floods its banks. But Sirius still shines out as before, the brightest star in the northern sky, calm and unmoved.

The Egyptians were not alone in looking to the heavens to find order. Around 500 BC the followers of the Greek mathematician Pythagoras, having discovered the mathematical laws of musical pitch, inferred that planetary motions produce a 'music of the spheres', and developed a 'therapy through music' to bring humanity into harmony with the celestial spheres. Aristotle developed the idea that the realm of the stars was a state of eternal and unchanging perfection. The heavenly bodies were unblemished globes composed of a fifth element or essence, different from the four terrestrial elements of air, fire, earth and water. The Sun, Moon, planets and stars were carried round on spheres nestling inside each other. The medieval Christian Church took over many of Aristotle's philosophical ideas, and his idealised view of astronomy (refined subsequently by the observations of Ptolemy) became religious dogma.

It has been one of the triumphs of modern astronomy to show that the realm of the stars, far from being a constant and tranquil region, serene with celestial music, is in fact a raging inferno. Here there is incredible violence: massive explosions, appalling eruptions and fury, outpourings of searing energy on a scale unimaginable on Earth. And there *is* change, for even the stars die. Sirius will not shine for ever.

The other great achievement of modern astronomy has been to show humanity how low it ranks on the scale of things in this universe. Once, it had seemed obvious that the Earth was the centre of the universe and that the habitation of humanity was the point around which everything else revolved. Although some Greek astronomers realised that the solar system had to be bigger than most people imagined possible, the conventional view remained that the Earth was surrounded by a set of crystal spheres, not far distant, which carried tiny pinpoints of light. The stars and the planets appeared to be small because they *were* small, not because they were far away.

In the Christian era this view of the natural order of things became coupled to the theological view of the way the world was ordered. There was a satisfying harmony between the world as revealed to the senses – with the stars and planets revolving around the Earth – and the world as revealed by religious faith, with humanity as God's special creation having dominion over the rest of the natural world.

In the year 1610 this satisfying picture was shattered when Galileo Galilei, the Italian astronomer-physicist, published the results of observations he had made with one of the first telescopes. But even then astronomers still thought that the cosmos was a pretty small place, hardly larger than the solar system. It grew a bit over the next three centuries, as astronomers realised that the stars were other suns like our own, populating the galaxy or star city that our culture calls the Milky Way. But it was not until the 1920s that astronomers discovered the true, shocking scale of the universe. The man largely responsible for this twentieth-century revolution in humanity's understanding of its true status was Edwin P. Hubble, the astronomer after whom the Hubble Space Telescope was named.

Humanity, it turns out, has an insignificant place in the great scheme of things. The human race lives not in the New York, nor the London, nor the Paris of the universe, but in its suburbs, in the attic of an unremarkable house down a dull little cul-de-sac. Far from being at the bustling centre, the Earth is but one of nine planets revolving around a perfectly ordinary star two-thirds of the way out on one arm of a perfectly ordinary spiral galaxy, the Milky Way.

For us, the Sun – our own star – is everything, the giver and sustainer of life, the essence of our continued existence. Yet the Sun is only one among many stars in the Milky Way. Indeed, there are so many stars in the Milky Way that an entire human lifetime would not suffice to count them. A rough estimate is that there are 100 billion stars in our galaxy; that is the number of grains of rice that could be packed into St Peter's Basilica in Rome.

And our galaxy, the Milky Way, is only one member of a group of thirty or so galaxies in the 'local' vicinity. Further away in space, astronomers have identified clusters containing tens of thousands of galaxies each. No one can count how many galaxies there might be in the universe, but pictures from the Hubble Space Telescope have led astronomers to quadruple their estimate, from 50 billion to 200 billion. This would mean that in total there are more than 20,000,000,000,000,000,000,000 stars – 20,000 billion billion – in the universe. These are staggering numbers, incomprehensible in their scale.

The stars have always prompted the big questions. At times, as in Ancient Egypt and medieval Europe, they have appeared to sustain religious conviction; at other times, especially since Galileo, they have made orthodox religious belief psychologically more difficult to maintain. But faced with the newly revealed scale of the universe, the questions have never loomed quite so large as they do today. How did the universe begin? How will it end? What is the true nature of time? Are we alone?

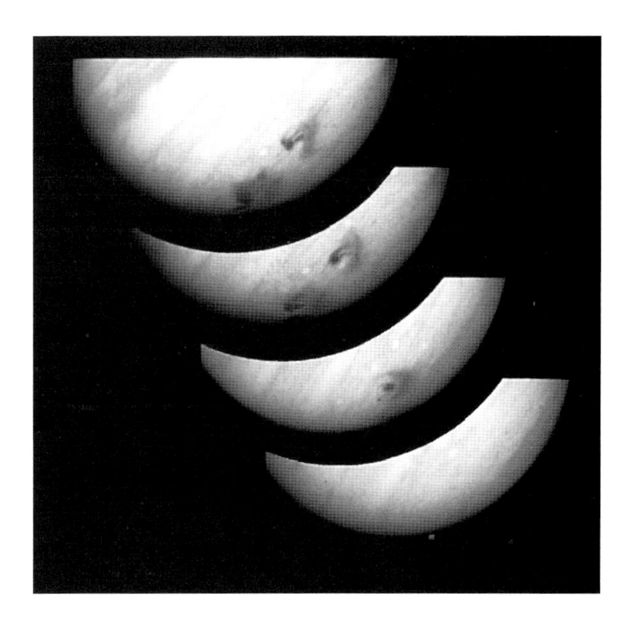

Watching the effect of Comet Shoemaker-Levy crashing into Jupiter in July 1994. The comet had already broken up into fragments and these images show the evolution of some of the impact sites (from lower left to upper right) starting when the plume or splash created by the impact was clearly visible. The effect of winds and of further fragments hitting the planet can be seen in the later images, with the last taken five days after the first.

Credit: R. Evans, J. Trauger, H. Hammel, the HST Comet Science Team and NASA.

The modern picture of the universe has been patiently and painstakingly built up by astronomers over the years. Often, the evidence upon which modern theories are built is esoteric and incomprehensible to lay people, for whom seeing is believing. Today, however, they can see the evidence. And for that, we can thank the Hubble Space Telescope, orbiting 375 miles (600 kilometres) above our planet's surface.

The pictures sent back from Hubble have, really for the first time, brought home to ordinary people the immensity of the forces at work in the heavens. We have seen the effect of comets smashing into the surface of Jupiter with an explosive

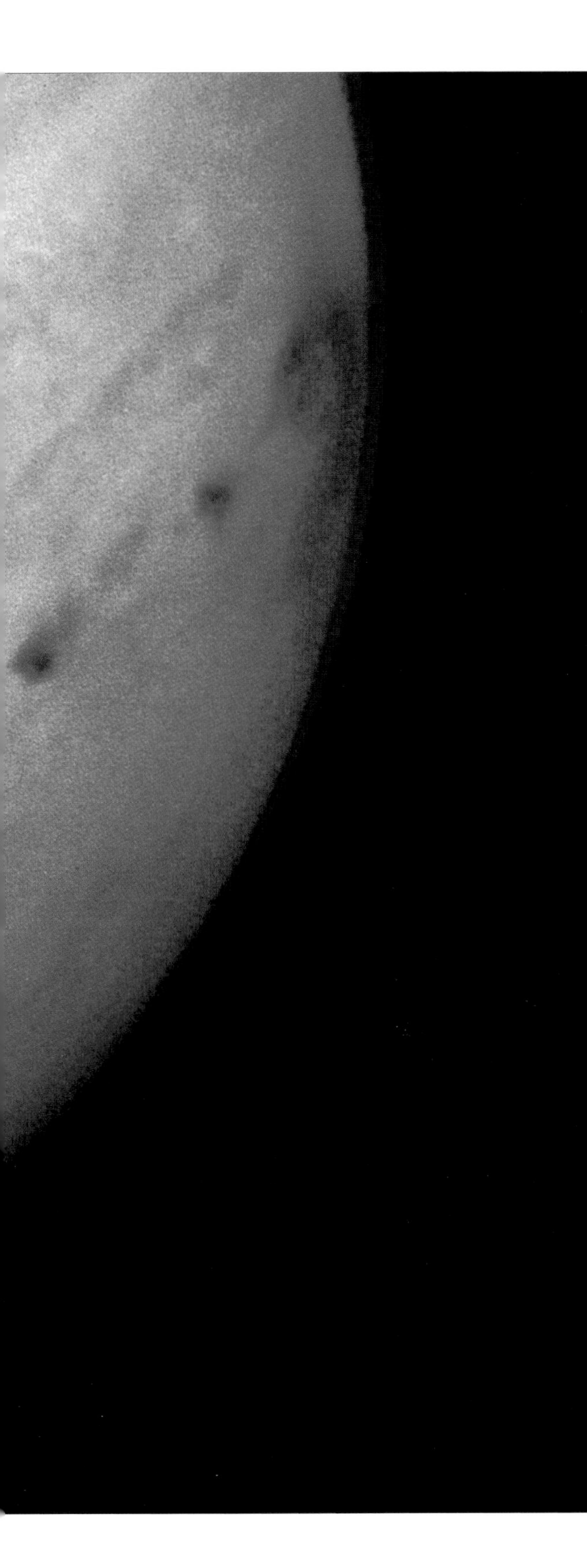

power of 100 million megatons (the actual impact site itself was hidden just round the back of the planet); watched storms brewing on the surface of Saturn; witnessed doomed stars exploding amid brilliant firework displays, with loops and rings of gas light-years across surging outwards into space; black holes, white dwarfs, red giants . . . all have been revealed.

Perhaps most astonishing of all have been the pictures showing the birth of stars. Focusing on the Eagle nebula in the constellation Serpens, 42 million billion miles (67 million billion kilometres) away from Earth, Hubble's Wide Field and Planetary Camera beamed back to us a striking image of vast columns of cool gas and dust, which are dense enough to collapse under their own weight to form young stars. As they examined the pictures, astronomers realised that they were watching an interstellar 'hurricane', in which the 'wind' was actually a blast of ultraviolet light from nearby hot young stars, heating the gas along the surface of the columns and boiling it off into space to reveal the newly born stars beneath.

These events in the Eagle nebula took place 7,000 years ago, more than 1,000 years before the first Pharaoh ascended the throne of Egypt. Although light is the fastest thing in the universe, travelling at about 186,000 miles (300,000 kilometres) per second , it has taken 7,000 years to cross the void between the nebula and the Earth. By recording these images, the Hubble Space Telescope is acting as a time machine – it can actually show us the past. Astronomers realised that by searching for

The sites at which fragments of Comet Shoemaker-Levy 9 hit Jupiter in 1994 are visible as angry-looking spots on the planet's surface. Eight impact sites are visible to the trained eye – all of them changed rapidly over a period of just a few days.

Credit: Hubble Space Telescope Comet Team and NASA

Above: The furthest ever peek to the outermost limits of the observable universe has revealed thousands of galaxies never before seen. This is the famous 'Hubble Deep Field' (only a quarter of which is shown here). The unusual and unexpected shapes of the galaxies have yielded important information about the early history and evolution of the universe. Some of the galaxies may have formed less than one billion years after the Big Bang.

Credit: Robert Williams and the Hubble Deep Field Team (STScI) and NASA

the most distant objects, whose images had taken the longest to reach us, Hubble would allow us to look back at the very beginnings of the universe: the beginning, for us, of time.

At the personal initiative of Dr Robert Williams, the then director of the Space Telescope Science Institute at Johns Hopkins University in Baltimore, Hubble was set to look for the faintest objects in one small patch of sky. In scientific terms, this was a risky experiment: it required an exposure ten days long towards the end of 1995, at a time when scientists around the world were clamouring to get to use the telescope for just a few seconds. (That demand for observing time remains so great that today nine out of ten proposals to use the telescope have to be turned down.)

But the payoff from the so-called 'Deep Field' search was spectacular. As the telescope peered out to the edge of the visible universe it began to send back pictures formed from events whose images had been travelling at the speed for light for billions and billions of years; events soon after the moment of creation. The pictures revealed new, young galaxies – great celestial star cities – in the very act of formation.

Astronomers processed 342 frames photographed between 18 and 28 December to form a single image of deep space. When the Hubble Deep Field image was released to a meeting of the American Astronomical Society in January 1996 it created a scientific sensation. Light from the closest of the galaxies in the picture (which tend to be the brightest) had taken about 5 billion years to reach the Earth. The most distant tend also to be the dimmest and most of the galaxies are so faint – about 4 billion times fainter than can be seen with the human eye – that they had never been seen even with the largest earthbound telescopes. Light from the most distant galaxies had set off more than 10 billion years ago – possibly as long as 14 billion years ago – around the time when the universe was formed. The astonished scientists crowding around copies of the photograph saw some 1,500 galaxies, each containing up to 100 billion stars, in various stages of formation. There were ellipses and spirals; some shaped like beach balls and footballs; others, long cigar-shaped clusters of stars.

The picture stunned the scientists because it challenged accepted theory: before they saw the Deep Field, many astronomers had believed that at the time the images were generated the universe would have been too young for galaxies to have formed. As so often, evidence gathered by the Hubble Space Telescope forced them to rethink. The telescope offers more than just snaps, more than just space tourism; it is both a precise and a provocative scientific instrument.

Technology has been at the heart of the revolution in astronomy. Galileo started it, when he constructed his own telescope at the beginning of the seventeenth century. Some sixty years later, Isaac Newton – best known for his mathematical work on gravity and motion – invented the first reflecting telescope and thus set the pattern for the most powerful astronomical telescopes since. The mirror in Newton's telescope measured 1 inch (2.5 centimetres) across. Edwin Hubble realised the great scale of the universe in the 1920s because he had access to the best technology, the biggest telescope then in existence: the 100-inch (254-centimetre) reflector on Mount Wilson near Pasadena, California.

After the Second World War there was a glut of cheap radar and communications equipment available and astronomers began to build experimental telescopes that 'listened' for radio emissions from stars rather than 'looking' at their light in the visible part of the spectrum. (In fact, Karl Jansky, an American radio engineer, had been the first to detect radio waves from the Milky Way – in 1931 – when doing research into static for the Bell Telephone Laboratories. But lack of interest and the outbreak of the war led to Jansky's discovery being largely ignored.) To their surprise, astronomers discovered that many different objects emitted large amounts of energy at radio wavelengths. This led to a flourishing of radio astronomy in the post-war years and a succession of startling discoveries, ranging from the discovery that Jupiter emits radio

Three selected areas of the sky taken from the Hubble Deep Field. The pictures are the result of putting together many separate frames (342 were taken in all and 276 have been used in this picture). This is a real-colour image formed by superimposing separate exposures taken in blue, red, and infra-red light. The colours reveal information about the distance, age and composition of galaxies.

Credit: R. Williams, the Deep Field Team (STScI) and NASA

waves when its moon Io passes across the face of the planet, to the 'echo' of the Big Bang which started the whole universe off. These all but eclipsed optical astronomy.

But *all* telescopes face a problem which earth-bound technology cannot solve. Their 'view' of the stars is distorted or totally obscured by the Earth's atmosphere. Some radio wavelengths are completely absorbed. Short wavelength emissions – ultraviolet, X-ray and gamma-ray – produced by the most energetic and violent of cosmic events are blocked.

Optical telescopes are hampered. To us, the atmosphere appears clear; yet the soup of gases which we live in and breathe interferes with light. Astronomers are rather like minnows swimming in a lake or river, trying to understand what is happening on the riverbank when all they can see is distorted by the water, by ripples on the surface, and by floating debris. Light from distant galaxies that has travelled intact for millions of years across the vastness of space is at the last moment garbled by the turbulence of the Earth's atmosphere just above the waiting apertures of the astronomers' telescopes.

To try to get a clearer picture, astronomers have dragged their equipment to some of the remotest locations in the world. Particularly favoured locations are isolated hilltops, some on islands in the middle of an ocean. There the air is less turbulent and the deterioration of image quality is reduced. So in the 1970s the scientific nations of Europe started to move their telescopes from ground-level sites (where they were threatened also by 'light pollution' from city street lamps lighting up the sky and blotting out faint stars and galaxies) to the top of La Palma, one of the Canary Islands in the Atlantic Ocean. At around the same time, American and European astronomers began to place telescopes on the extinct volcano of Mauna Kea – the White Mountain of Hawaii. This site is so high that some astronomers suffer altitude sickness when tending their telescopes.

Perversely, many of the characteristics of the Earth's atmosphere that ensure that it sustains life are also the ones that cause astronomers most difficulty. Water vapour, for example, absorbs longer-wavelength infra-red (or heat radiation), so preventing sea-level astronomers from studying this type of output from the stars. One great advantage of Mauna Kea is that it is so high as to be above much of the water vapour in the atmosphere, and today the mountain top is dotted with infra-red telescopes.

The other great protector of life on Earth, the ozone layer, is also a source of astronomical grief, because ozone absorbs ultraviolet light. As the ozone is located high in the stratosphere, it is no help climbing to the top of Mauna Kea this time.

Ultraviolet, infra-red, visible light, X-rays and gamma-rays: for all of these, there is only one answer. On a clear day, you can see for ever – so long as you go into space itself.

The pioneers of rocketry and space travel discussed the theoretical advantages of putting telescopes into orbit around Earth more than seventy years ago. But serious calculations began only after the war when, in 1946, the American Lyman Spitzer was the first astronomer to show the advantages of orbiting observatories. With the only rocket technology being commandeered German V2s, his was something of a lone voice in astronomy until the 1960s, with the advent of NASA and the demonstration – through the Mercury, Gemini, and Apollo programmes – that America had 'the right stuff' for space exploration.

But even before it was possible to send satellites into Earth orbit, many rockets were launched for brief sub-orbital flights with astronomical instruments on board. X-ray astronomy, for example, started as early as 1949 when a group of researchers at the US Naval Research Laboratory using V2 rockets discovered that the Sun emitted X-rays. These pioneering rocket flights were succeeded in the 1960s by satellites sent up to observe the heavens at wavelengths invisible to earthbound astronomers: infra-red; ultraviolet; gamma-ray and X-ray. These orbiting observatories led to the rapid development of astronomy in these new regions of the spectrum. But the creation of a large optical telescope above the clouds was longer in coming to fruition.

The idea gained credibility with the demonstrable sophistication of space hardware in the 1960s. But, after the excitement of the Apollo Moon landings in the late 1960s – and under the impact of economic recession triggered by the rise in oil prices in the early 1970s – America's headlong enthusiasm for space exploration began to wane.

NASA presented its first plans for an orbiting optical observatory to Congress in 1974. The scheme envisaged a reflecting telescope with a primary mirror 10 feet (3 metres) across, which would be brought back down to Earth periodically for routine maintenance. The plan was rejected as being too expensive.

The nations of Europe had had an unhappy experience with space exploration. Some – notably the French and the British – had tried and failed to go it alone. But European-wide collaboration in space was scarcely more successful. Two rival organisations failed: the European Space Research Organisation (ESRO) and the European Organisation for the Development and Construction of Space Vehicle Launchers (ELDO). Eventually, in May 1975, they clubbed together into the European Space Agency (ESA), headquartered in France and with a heavy French dominance of the Ariane launcher technology.

NASA got its new European counterpart interested in the space telescope project with the Europeans providing some of the imaging technology and meeting some of the costs. With persistence and international co-operation, in 1977 NASA won the political support it needed to start building the telescope. Support was forthcoming partly because NASA had scaled back its plans. The mirror was to be smaller – 8 feet (2.4 metres) across – significantly reducing the costs of the project. The general configuration of the scientific instruments was also changed to reduce the bulk of the telescope and thus make it cheaper. The idea of returning to Earth for maintenance was abandoned; the telescope would be designed for spaceside maintenance by visiting astronauts carried by America's Space Shuttle. Nevertheless, as the newly funded project gradually took shape, it became clear that it would be arguably the most sophisticated scientific satellite ever built. And its name: the Hubble Space Telescope.

Astronomy, in the popular imagination, is the pursuit of absent-minded elderly men with wild white hair. No one could be more different from this stereotype than the century's greatest observational astronomer, Edwin Powell Hubble. Born in 1889, the son of an insurance clerk in Marshfield, Missouri, he grew up tall, handsome and athletic, and won a scholarship to study law at the University of Chicago. He did take some science and technology courses as well, but in 1910 he won a Rhodes Scholarship to Oxford to study jurisprudence.

Oxford turned Hubble into the very model of an English gentleman: he won a half-blue for athletics; he spoke with an English accent; and he affected a Dunhill briar pipe. Thirty years after

Opposite: The Space Shuttle Discovery carries the Hubble Space Telescope into space on 24 April 1990. It weighed 11.25 tonnes and cost more than $1.5 billion.

Credit: NASA

Oxford, his affection for all things British showed through undiminished when he opposed American isolationism at the beginning of the Second World War and actively campaigned for the USA to side with the British.

Nevertheless, after Oxford Hubble returned to America and took up astronomy. In the very month the First World War broke out in Europe he started studying for a PhD at Chicago University's Yerkes Observatory. His studies completed, he joined the US infantry in 1917, and did not resume his work until 1919, when he joined the staff at the Mount Wilson Observatory in California. By happy coincidence, the biggest telescope in the world at that time, the 100-inch (254-centimetre) Hooker reflector, was ready for use just as he arrived.

Over the next decade Hubble's meticulous measurements started a revolution in humanity's view of the universe. This tall, vigorous, immensely confident man – who would direct the movement of the massive telescope with his pipe clenched firmly between his teeth – looked nothing like the conventional picture of the brilliant but batty scientist. However, his results overthrew calculations made by Albert Einstein, the century's most eminent mathematical scientist (and one who certainly looked the part).

First, in 1924 Hubble revolutionised astronomers' ideas of the size of the universe by showing that some of the nebulae visible in the night sky were not simply hazy patches of light in our own Milky Way, as had been assumed, but were actually immensely distant galaxies in their own right. He made this discovery by taking advantage of the properties of a special type of star, known as the 'Cepheid variables'. These stars pulsate in a regular manner, and the rate of pulsation allows astronomers to work out how brightly the star is shining. They then compare this with the stars' apparent brightness, after the light has travelled all the way to Earth. The difference tells them how far the stars are away. Hubble studied thirty Cepheid variables in the Andromeda nebula and found that they were so faint they had to be hundreds of thousands of light years away, and therefore so was the Andromeda nebula itself. Time and again he found the same sort of result; the nebulae were simply too far away to be inside our own galaxy. Hubble himself was cautious, never referring to them as galaxies but as 'extra-galactic nebulae'. But despite some resistance by die-hard scientific conservatives, acceptance quickly grew that the universe was much larger than anyone had previously thought, and that our galaxy was only one among a huge number of celestial star cities.

Opposite: Free from the obscuring veil of the Earth's atmosphere, the Hubble Space Telescope now has a near-perfect view of the universe in which we live.

Credit: NASA

Five years later Hubble produced another sensation: he proved not only that there were other galaxies beyond our own, but also that they were moving still further apart, and that the universe as a whole was not static but expanding all the time. This finding knocked away the last prop for any sort of belief in an orderly and static realm of the stars; here was a cosmos perpetually on the move.

Hubble had been studying the light produced by distant galaxies. As well as straightforwardly photographing the light coming from distant stars, astronomers pass it through a prism to split the white light up into its component colours. They do this because different chemical elements emit different colours when they incandesce, and so this splitting of the light into a spectrum in effect allows astronomers to uncover the elements contained in any particular star. In 1914 the American astro-nomer Vesto Slipher, working at the Lowell Obser-vatory in Arizona, had discovered that the spectra of light from the nebulae were all shifted from their expected positions towards the red end of the spectrum – a phenomenon known as 'redshifting'.

Redshifts are similar in principle to the Doppler effect. Anyone who has stood on a railway platform and listened to the horn of a passing train will understand the effect. As the train approaches its horn is higher pitched than normal, and as soon as it passes there is a distinct drop in pitch. Colour is to light as pitch is to sound, and redshifted light is the equivalent of the lower pitch of the receding train: the star too must be receding. Hubble also found that, wherever he looked, the redshift was more marked among fainter, more distant galaxies.

By 1929 Hubble announced his discovery that the redshift in a galaxy's light is directly proportional to its distance from Earth. In other words, the more distant a galaxy, the faster it is moving away from us; which means that the entire universe is expanding at a constant rate in all directions. (The expansion of the universe means that the redshifts in galactic light are not truly from the Doppler effect. A Doppler redshift results from movement through space, whereas the cosmological redshift stems from the expansion of space itself. The best way to imagine this is to think of space as a large balloon with the galaxies stuck on the surface like paper dots. As the balloon is inflated the dots move away from each other as the surface of the balloon expands – even though each dot is firmly stuck to its spot on the balloon's surface.)

Hubble's revelations came as a profound shock to his contemporaries. Albert Einstein had formulated his General Theory of Relativity in 1915 but Einstein (along with everyone else) was so wedded to the idea of a static universe that he slipped an extra 'fudge factor' into his equations when they started to suggest an expanding universe. Hubble showed that general relativity was correct, and that Einstein himself had misunderstood its implications. It was, Einstein later confessed, the biggest mistake of his career.

Hubble remained active in research nearly until his death in 1953, shortly after helping to complete the installation of a 200-inch (508-centimetre) telescope on Mount Palomar in California. This extraordinary man, having acquired a rich wife, moved among the stars in quite another way. The Hubbles were key figures in the social world of Hollywood during the 1930s and 1940s: Walt Disney, Charlie Chaplin, the Marx brothers, Aldous Huxley and the newspaper magnate William Randolph Hearst were all friends of the astronomer. Hubble left instructions that there should be no monuments to him. There was neither a formal funeral nor a memorial service; Edwin Hubble lies buried in an unmarked grave in an unknown spot.

Today, the Hubble Space Telescope is a far better monument to the astronomer than any stone statue or brass plaque. However, its path to the stars was not an easy one.

Having won approval for the telescope in 1977, NASA ran into trouble when the Space Shuttle Challenger exploded in 1986 and all Shuttle flights were suspended. But the disaster only increased NASA's determination to make the telescope project happen; the agency, increasingly beleaguered by budget cuts, needed the project to succeed for the sake of its own reputation. The Hubble Space Telescope was officially designated as the first out of four missions in NASA's 'Great Observatories' programme. Hubble was followed by the Compton Gamma-Ray Observatory launched in 1991, then there was supposed to be an X-ray satellite and the programme would culminate in the Space Infra-Red Telescope facility due for launch in 1999. But NASA's priorities have changed in the interim and it is not clear if the Great Observatories programme will be seen through to completion.

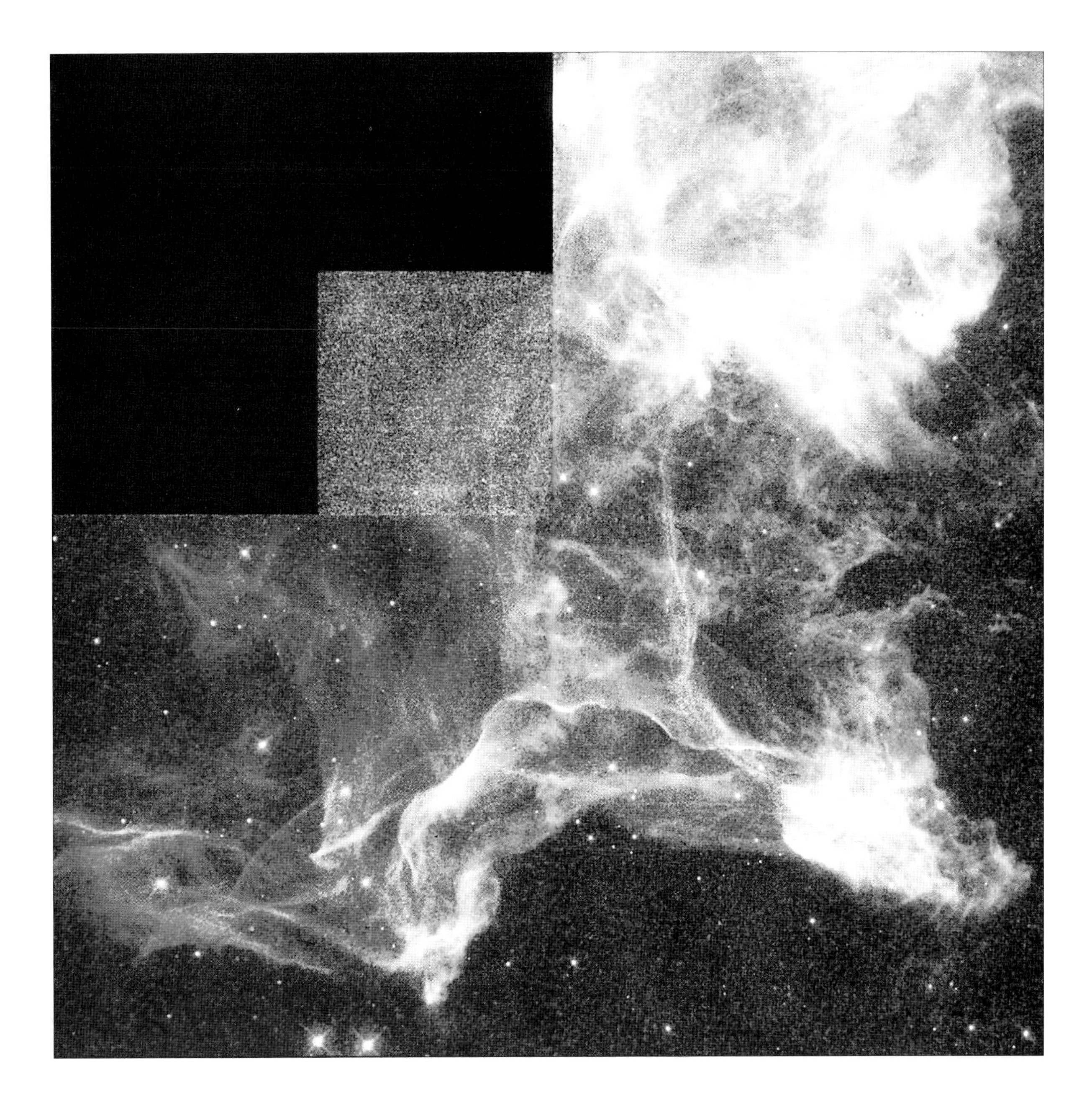

The Hubble Space Telescope that was eventually loaded into the cargo bay of the Space Shuttle Discovery was a gargantuan piece of technology. It weighed 11.25 tonnes, was the size of a city bus, and its total cost was suitably astronomical, having soared beyond $1.5 billion. On 24 April 1990 it was launched into space, carrying with it the hopes of the world's astronomers.

Once released from Discovery's cargo bay into

This image shows a nebula called the Cygnus Loop. Covering a region on the sky six times the diameter of the full Moon, the Cygnus Loop may look beautiful, but it is actually the expanding blastwave from a stellar cataclysm. Some 15,000 years ago, a dying star blew itself up in a supernova explosion, scattering dust and gas into interstellar space. The Cygnus Loop remains as the most visible sign of that destruction. This supernova remnant lies 2,500 light-years away in the constellation Cygnus the Swan.

Credit: NASA

Above: The first image to demonstrate, on 18 December 1993, that the repair mission had been a success. The star is Melnick 34, in the 30 Doradus nebula (also known as the Tarantula nebula). The image on the left is one of the best achievable from the ground – taken at the European Southern Observatory, high in the Chilean Andes. The central image is from the HST before the repair; but on the right, after the repair, not only the central star but faint background stars are also clear and sharp.

Credits: ESO and WFPC2 Team and NASA.

Opposite: Hubble has revealed for the first time the aurorae that encircle the poles of the planet Saturn. These have never been seen from Earth, whose atmosphere prevents the ultraviolet light from reaching ground-based telescopes. The spectacular lightshow results from the battle between the planet's magnetic fields and the flow of electrically charged particles coming from the Sun. The curtains of light rise more than a thousand miles above Saturn's cloud-tops.

Credit: J. T. Trauger (Jet Propulsion Laboratory), and NASA

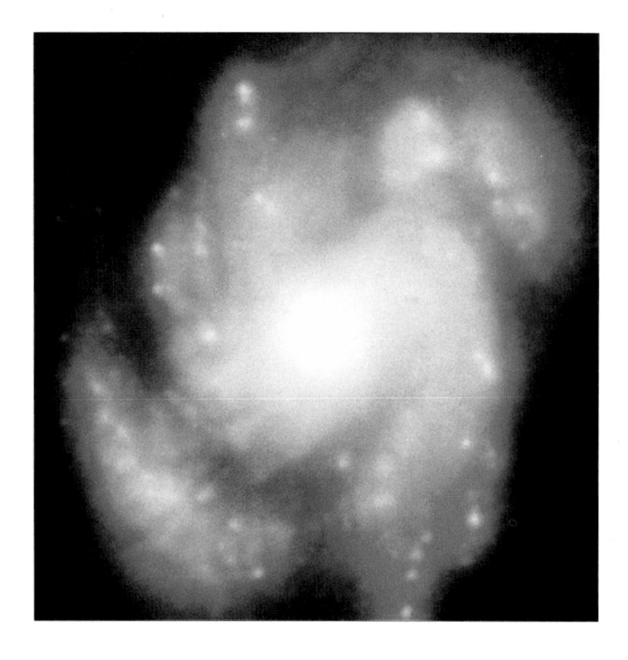

The true extent of the defect in the mirror of the HST is evident in these two images of the nucleus of a galaxy, M100, lying several tens of millions of light years distant. On the left is the original Hubble image taken with the instrument known as the Wide Field and Planetary Camera. Although some stars can be discerned, the detail is poor.

On the right is the replacement instrument's picture, taken with corrected optics. The image is crisp and clear and shows lots of detail. The galaxy M100 is one of the brightest members of the Virgo Cluster of galaxies. The galaxy is in the spring constellation of Coma Berenices.

Credit: NASA

its own orbit 375 miles (600 kilometres) above the Earth's surface, Hubble went through various engineering stages to prepare it for its scientific mission. Power was provided by solar panels 40 feet (12 metres) long, which had to unfasten themselves and face the Sun; communication had to be established using Tracking Data Relay Satellites – a sort of orbital telephone exchange cum postbox.

These initial preparations were accompanied by the sort of glitches that can be expected on any

The G1 'globular cluster' of stars (also known as Mayall II) is the large, bright ball of light in the centre of the picture. It consists of at least 300,000 old stars and orbits the Andromeda galaxy (M31), the nearest major spiral galaxy to our Milky Way. Close study of the data beamed back by the Hubble Space Telescope permits astronomers to see the fainter helium-burning stars in G1 where helium rather than hydrogen gas is being fused together.

Credit: M. Rich, K. Mighell, J.D. Neill (Columbia University), and W. Freedman (Carnegie Observatories) and NASA

project of this complexity. NASA's engineers discovered that a cable was blocking rotation of an antenna, but they corrected that by limiting the antenna movement. Then they realised that the telescope was pointing at the wrong parts of the sky, so they changed the software in the controlling computer.

In late May, just a month after launch, the telescope beamed back the first test pictures, of the open star cluster NGC 3532, which is about 1,260 light-years from Earth. They were greeted with cheers in the NASA control centre at Greenbelt, Maryland. But the applause died away when professional astronomers started to examine the photographs.

With remarkable prescience, Professor Alec Boksenberg, then director of Britain's Royal Greenwich Observatory, remarked when he saw the pictures that while the telescope 'is making remarkably good progress, given the complexity of the whole system . . . the whole telescope is not properly in focus yet.' On 27 June NASA could conceal the size of the disaster no longer: Hubble could not focus properly. Not only that, it had a bad case of the shakes. At first, NASA suspected that one of the mirrors in the complicated system of optics must have been jarred out of position during launch. The large primary mirror, 8 feet (2.4 metres) in diameter, collects light from distant objects and focuses it on a smaller secondary mirror, which in turn directs the focused images into the scientific instruments carried on board. NASA staff thought that the secondary mirror had been knocked out of its proper position, producing unfocused images. They spent two months fiddling with the mirror's position, using radio signals to make minute adjustments, before they finally admitted defeat, and acknowledged that there was what they called a 'spherical aberration'. In plain language the primary mirror, the very heart of the telescope that had cost so much money and taken so much time to put into space, was the wrong shape.

It was a period of near-despair. NASA set up an official investigation into what had gone wrong, headed by Lew Allen, director of the agency's Jet Propulsion Laboratory in Pasadena, California. What finally emerged as the twin causes were somehow very human: budgets and secrecy.

The company that made the telescope's mirrors, for $450 million, was the Hughes Danbury Optical Company, then a division of Perkin-Elmer. Its first public statements revealed confusion about the cause of the error. Dr Jack Rehnberg, chief of the company's space science office, said it appeared 'Something inherently, fundamentally was not done right.' However, another spokesman for the company, Thomas Arconti, insisted that the mirrors on Hubble had been made precisely to NASA's specifications. Then Dr Rehnberg accepted that there was a possibility of human error in the way testing equipment had been arranged during the manufacture of the mirrors.

NASA's deputy manager for the Hubble project, Jean Olivier, confirmed that while the two mirrors had been tested individually prior to launch, they had never been tested as a combined pair, because that would have cost hundreds of millions of dollars. It then transpired that military secrecy had restricted NASA's ability to oversee the critical work of constructing the primary mirror. The technology involved in making the mirrors was similar to that used in spy satellites; these too use powerful telescopes, but pointed down at the Earth rather than out to the heavens. Perkin-Elmer had been given the contract because it had experience in making components for these military spy satellites. But because of its military connections, the plant was off-limits to all but a few NASA staff.

The investigation revealed that testing had not been thorough enough during manufacture, and a systematic error had crept in. In the end, the curvature of the mirror, 94.5 inches (2.4 metres) across, was wrong by about one-fiftieth of the diameter of a human hair. By 14 August NASA investigators had conducted preliminary tests on a 'null corrector' used to measure the mirror's surface during the grinding process, and found that one spacing in the instrument was a millimetre off specification. This was enough to create an error just two-millionths of

a metre in the curvature of the mirror itself (just less than one-ten thousandth of an inch).

However, although the mirror's highly polished configuration was the wrong one, it was nevertheless precise. Astronomers realised that the very regularity of the error would make it possible to devise correcting optics – space spectacles – to restore the required image quality. Hope was restored.

Meanwhile NASA had more worries. Two 40-foot-long (12-metre) solar panels provide power for the telescope; made of Teflon sheets covered with photovoltaic cells, the solar arrays convert the Sun's energy into electricity for the scientific instruments inside the telescope. The problem was that the panels were fluttering like wings whenever the telescope moved in and out of sunlight, and the wobbling this produced, sixteen times a day, was interfering with the telescope's alignment.

The answer, clearly, was to send in a repair crew. It took three years to complete the detailed planning for the rescue mission, which was launched on 2 December 1993. Three days later, the crew of the Space Shuttle Endeavour donned their suits and sallied out into space to start five days of repairs.

The most spectacular series of space walks seen to date were watched on television by millions, who marvelled at the sight of the tiny white-clad humans dwarfed both by the immensity of the void and the bulk of the telescope they were servicing, with humans and machines hurtling through space at 18,000 miles (29,000 kilometres) an hour. So demanding was the work schedule that two teams,

The triumph of human ingenuity and manual dexterity over the human errors made in the making of the Hubble Space Telescope. Astronaut Story Musgrave in the course of a spacewalk during the Hubble repair mission.

Credit: NASA

Above and right: Edwin P. Hubble has a strong claim to being the most influential and important astronomer of modern times. Just as Galileo allowed scientists to break away from the Earth-centred view of the universe, so Hubble's vision and deep insight allowed them to see beyond our own galaxy; to realise that ours is but one among countless galaxies in an ever-expanding universe.

Credit: Hale Observatories / Science Photo Library

each consisting of a pair of astronauts, were deployed, so that when one team tired, the other could continue the work. The sequence of events had been meticulously planned and re-hearsed by the astronauts underwater (to simulate zero gravity conditions) at Houston and Huntsville, Alabama. But once they got to the telescope, they encountered unforeseen problems and had to improvise. The latches on the doors of some equipment bays would not close; one of the solar panels was bent out of shape; the radio on astronaut Kathy Thornton's spacesuit did not work properly, and messages to her from ground control had to be relayed via her fellow astronaut Tom Akers; the new solar panels would not unfurl, forcing the astronauts to crank them out by hand.

The delicacy of what the astronauts had to do can be illustrated by just one space walk, when Story Musgrave and Jeff Hoffman replaced the Wide Field and Planetary Camera, weighing in at 595 pounds (270 kilograms) and one of the main scientific instruments on the telescope. The new version of the camera had a coin-size mirror ground to a configuration that would exactly compensate for the flaw in Hubble's main mirror. In addition, the new camera was designed to 'see' ultraviolet light. The replacement operation was fraught with tension. The camera was supposed to slide into the telescope like a drawer. But the astronauts first had to remove the protective cover without touching the mirror itself. If, encumbered by spacesuits and gloves, they had touched the mirror, they would have knocked it out of alignment and contaminated its surface. They handled the operation flawlessly.

The mission had been completely successful. Hubble was able to 'see' with the precision its designers had originally intended, and the images of human heroism were supplanted by a flood of hauntingly beautiful cosmic pictures. The most distant parts of the universe sprang into focus; so sensitive was the refurbished Hubble that it was able to pick out incredibly faint objects. Imagine trying to see a 100-watt light bulb shining from as far away as the Moon (239,000 miles/385,000 kilometres), and you begin to get an idea of the power of the telescope.

Experiment followed experiment, with the results flowing thick and fast. Many results, as will be revealed in later chapters, confirmed astronomers' theories. Others did not, leaving scientists with fresh problems to solve. Above all, evidence gathered by Hubble suggested that the universe might be half the age that had been thought: 8 billion years rather than the conventional value of 15 to 16 billion years. But this meant that the universe was actually younger than some of the stars it contained – an obvious and deep contradiction. The result was a source of both frustration and joy to astronomers: frustration at the apparent violation of their cherished theories, and joy at the thought that there is still so much to know. It led one astronomer to declare that, thanks to the Hubble Space Telescope, 'the universe is in crisis'.

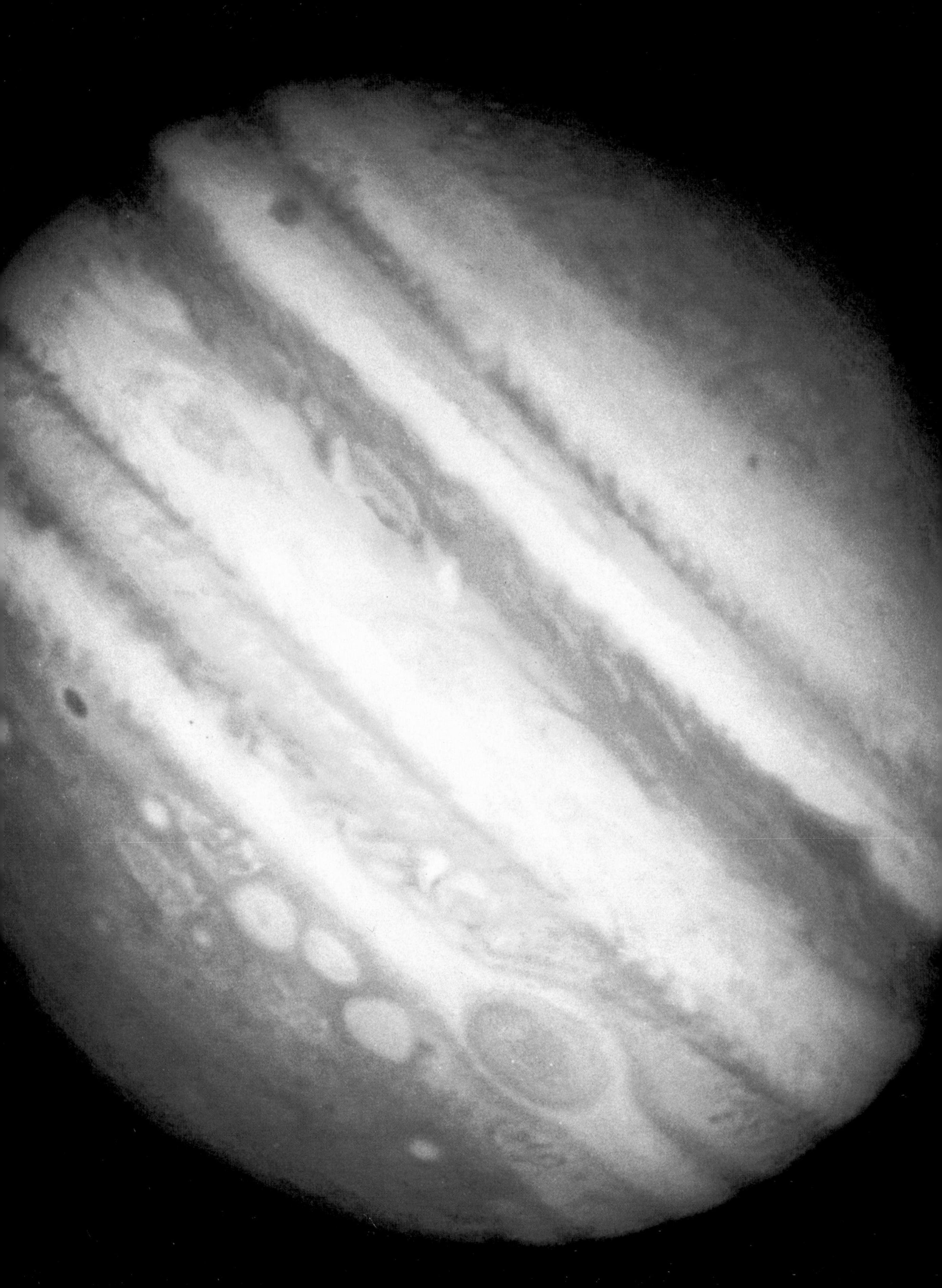

CHAPTER TWO

NEIGHBOURS

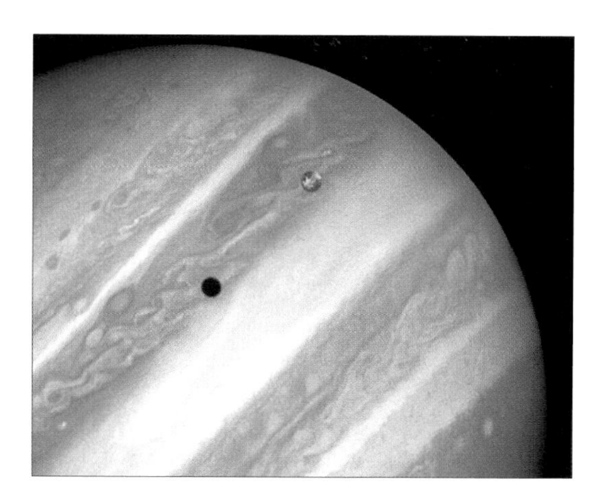

Humanity has always looked to the stars and imagined – hoped – that it was not alone. In ancient times our forebears looked at the constellations blazing out of the night skies and populated them with gods and heroes and beasts, drawn from the stories they told for entertainment and instruction, and drawn from their religion.

Opposite: A close-up of stormy weather systems on Jupiter, a gas giant and the largest planet in our solar system (indeed, more massive than all the other planets put together). This photograph clearly shows three white storms that are approaching the famous Great Red Spot, itself a storm that has been observed raging for more than 300 years. Two of the white storms were first seen forming 60 years ago. At the heart of these turbulent clouds gases are rising, drawing fresh ammonia gas to the surface. When this gas reaches the bitterly cold cloud tops, it starts to freeze, creating white ice crystals. Close observations by Hubble have allowed scientists directing planetary probes to receive up-to-date weather reports, vital to the success of their operations.

Credit: Reta Beebe and Amy Simon (New Mexico State University), and NASA

Above: This image shows Jupiter's volcanic moon, Io, sweeping past the massive planet at a speed of 38,000 miles per hour (17 kilometres per second). Io's shadow is thrown starkly onto the surface of Jupiter's atmosphere; the shadow measures approximately 2,270 miles (3,640 kilometres) across. Io is one of the 'Galilean' moons, so called because they were discovered by Galileo in 1610.
The other Galilean moons are Europa, Ganymede and Callisto. Io, which is roughly the size of our own Moon, is remarkable for its violent volcanic activity, caused by the destructive effects of the planet's gravitational pull on the surface crust of the moon. The white patches visible on the satellite are areas of sulphur dioxide frost.

Credit: J. Spencer (Lowell Observatory), and NASA

Today, men and women look at the stars and wonder if there is life out there; whether Earth is a cosmic one-off, a solitary freak, or whether the universe teems with civilisations and life-forms waiting to be discovered (or to discover us). Whenever a scientist announces that this or that piece of research shows that life beyond Earth is more likely, most people feel a curious and thrilling excitement, mixed with a tiny touch of terror. Films and books about extraterrestrials and space adventure pour out and are snapped up, regardless of the fact that they routinely draw their storylines from the absurd opposite ends of the spectrum: the wicked aliens are coming to conquer and make slaves of us; the wonderful aliens are coming to share their wisdom with us.

Biologists have for a long time understood that the key to life (at least as we know it) is planets. Stars are too hot, space is too cold; planets are the only option. Stable, close enough to a friendly neighbouring star to be warmed up (but not so close that they get burned up), packed with chemicals and elements for manufacturing the complex molecules of biology, and preferably with a comforting blanket of atmospheric gas to provide nourishment for the life-forms below while at the same time fending off the cosmic bombardments of rays that make the near-vacuum of space such a dangerous place.

The astronomers' problem is that planets are extremely hard to find. By definition, they are located right next to incredibly bright objects – stars – whose light overwhelms the viewfinders of distant telescopes. Astronomers are instead forced to look for indirect evidence of planets, or to work out where they are most likely to occur. And to do that, they need to understand how planets come to be created in the first place.

Some of the most exciting discoveries have come from a close examination of one of the most familiar constellations in the sky at night: Orion. According to one version of the myths of ancient Greece, Orion was a son of the sea god Neptune. The mightiest of hunters, Orion had the hubris to declare that he could slay any creature on Earth. To punish him for his presumption, the goddess Juno then called a scorpion out of its hole to sting Orion on the heel and kill him. Diana the huntress pleaded his case with Jupiter, king of the Olympian gods, and Orion was placed in the starry firmament. (With a tact remarkable for the ancients, the scorpion was placed at the opposite side of the sky; when Orion is visible, Scorpio is nowhere in sight.)

Just below Alnilam, the central star of the Hunter's belt, lies a faint, fuzzy patch of light called the Orion nebula. The light from the nebula that astronomers see today set out on its journey to Earth some 1,500 years ago, just as Attila the Hun was leading his barbarian horsemen to attack Rome. The light from Orion has carried with it on its long journey information about one of the most marvellous of all astronomical phenomena: the birth of new solar systems, of suns and planets. The Hubble Space Telescope has allowed us to analyse that information in more detail and more depth than was ever possible before.

The world upon which we stand, and the Sun that shines upon us, have been in existence for a long time. They were formed at least 4.5 billion years ago. By contrast, the stars shining in the Orion nebula have been around for no more than a million years. Think of the Sun as a middle-aged man whose appearance does not change much, apart from the first almost imperceptible signs of ageing, whereas in astronomical terms, stars of the Orion nebula are mere babies, a couple of days old, more or less at the beginning of their being and of a rapid phase in their development. The far-seeing eye of Hubble has allowed us to peek into a star nursery, and to see how it must have been at the birth of our own solar system.

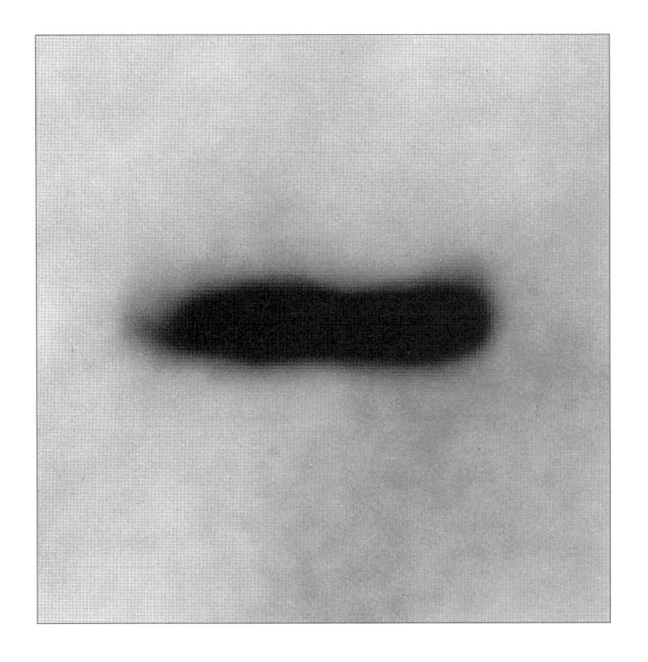

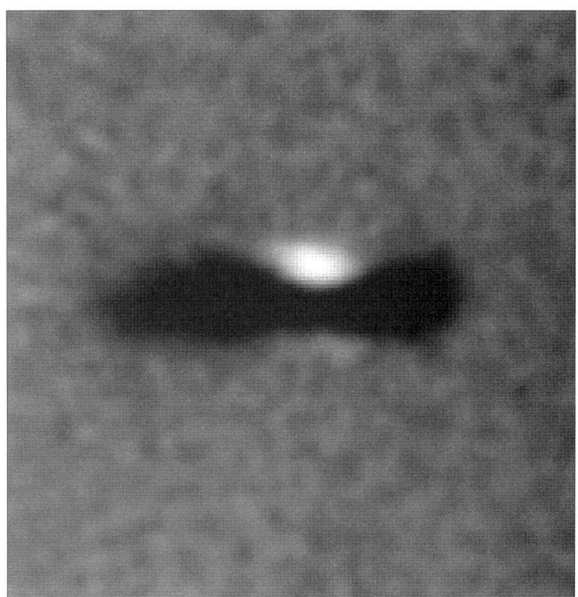

These two views of a protoplanetary disk (or 'proplyd') in the Orion nebula, seen edgeways-on, clearly show the flattened shape of the cloud of dust and gas swirling around a young star. The star itself is almost completely hidden by the nebula, although filters used to take the image on the right dim the glow from the nebula to allow its presence to become apparent.

Credit: Mark McCaughrean (Max-Planck-Institute for Astronomy), C. Robert O'Dell (Rice University), and NASA

In 1994 Dr Robert O'Dell and his colleagues from Rice University in Houston, Texas, used Hubble to take pictures of great clouds of dust and gas swirling around these young stars in the Orion nebula. They suspected that these clouds contained the raw materials for the formation of planets, and were looking for evidence that some of the clouds had the right geometric shapes to be the precursors of planetary systems.

The fundamental ideas underlying today's theories for planetary formation were first put forward by the German philosopher Immanuel Kant in 1775, and later elaborated by the French mathematician and astronomer Pierre-Simon de Laplace. They developed the 'nebular hypothesis' of planetary formation to account for the fact, easily observable in our own solar system, that the orbits of the planets almost all lie in the same plane. Basically, they realised that the orbits had to be the remnants of a primordial mass of small particles (the nebula or 'cloud') circling the Sun from which the planets grew by a process of the particles sticking to each other. Astronomers reasoned that if the particles were spinning around the Sun in a flat disk, like a huge pancake or plate, they would be moving in the same direction and at roughly the same speed, and quite close to each other, and would therefore have more of a chance of gradually drifting together through gravitational attraction. As the particles came together they would form clumps that would attract yet more particles, and would eventually become planets. These disks are known as protoplanetary disks ('proplyds' for short).

Some astronomers had suggested that the dust and gas clouds around the stars in the Orion nebula might merely be spherical shells quite unconnected to planetary formation. Crucially, the Hubble images clearly showed for the first time that the clouds were indeed disk-shaped, and were therefore prime candidates for proplyds. Some of the disks could just be made out by ground-based optical and radio telescopes, but their signals would have been interpreted as coming from stars themselves. Only Hubble's fine resolution (and its position above the Earth's atmosphere) showed these planetary nurseries for what they truly are. Even more exciting was the fact that Dr O'Dell's team found these disks all over the place; out of the 110 young stars they surveyed using Hubble, more than half – 56 in all – had proplyds. A year later, a fresh survey brought the tally up to 153. Moreover, as proplyds shine far less brightly than the stars, many more may have been overlooked even by Hubble.

The images provided striking evidence that the processes leading to the formation of planets may be common in the Milky Way. And if planets are common in the Milky Way, they are likely to be common in other galaxies. In short, it greatly increases the chances that planets – and therefore life – might exist beyond our solar system.

Detailed analysis of the images also confirmed astronomers' theoretical models about how planets are created. The particles of gas and dust drifting in great clouds between the stars slowly draw together by gravitational attraction. As a cloud shrinks in on itself, it starts to spin, gradually flattening out into a disk. At the centre, where the pressure is greatest, pairs of hydrogen atoms are forced to fuse together and are transmuted into single helium atoms, giving off great pulses of energy in the form of heat and light. In effect, the ball of hydrogen at the centre of the disk spontaneously ignites into a massive hydrogen bomb, creating a new star.

Over time, the dust and other debris swirling around this new star form small pellets which aggregate together into larger planetoids, which in turn come together to form large planets with stretches of empty space between them. In the inner parts of the system, close to the star's radiation, the heat boils off the remaining gas, leaving the planets as rocky lumps. In our solar system, examples of these planets are Mercury, Venus, Mars and of course Earth. Those planets further from the star are too far away for the radiation to have the same effect, and you end up with planets that are mostly gas, often huge, and with their rocky cores

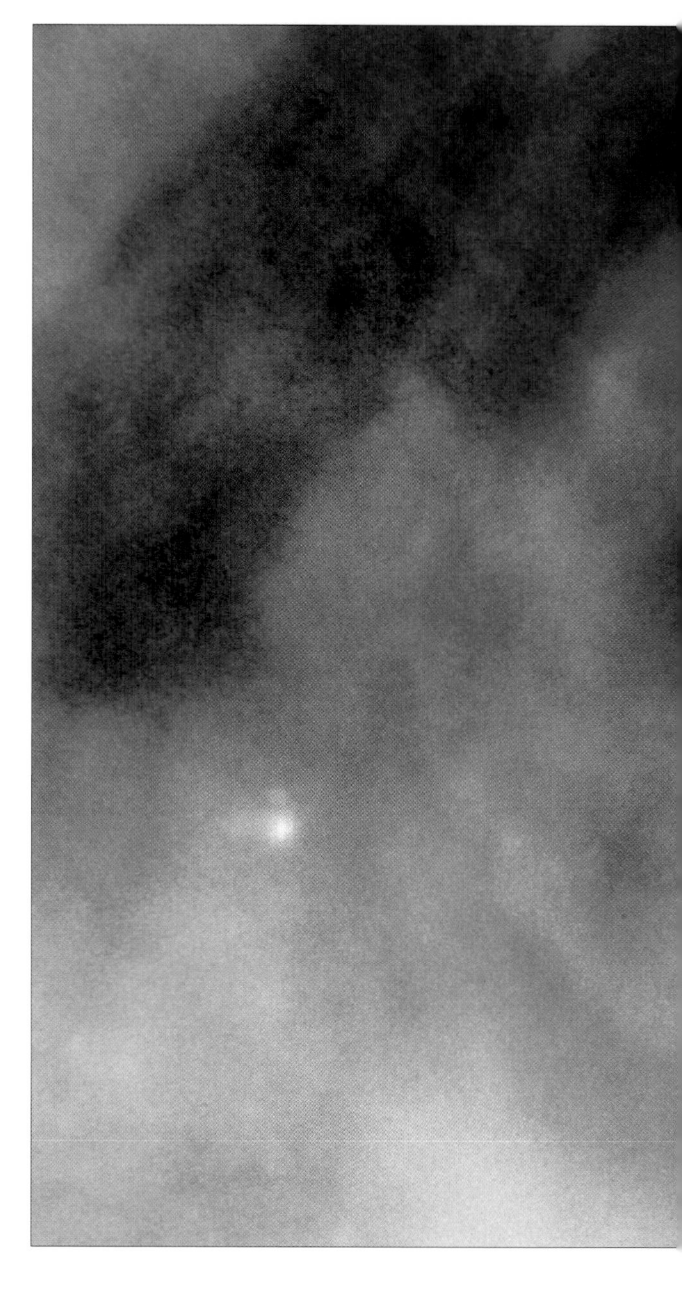

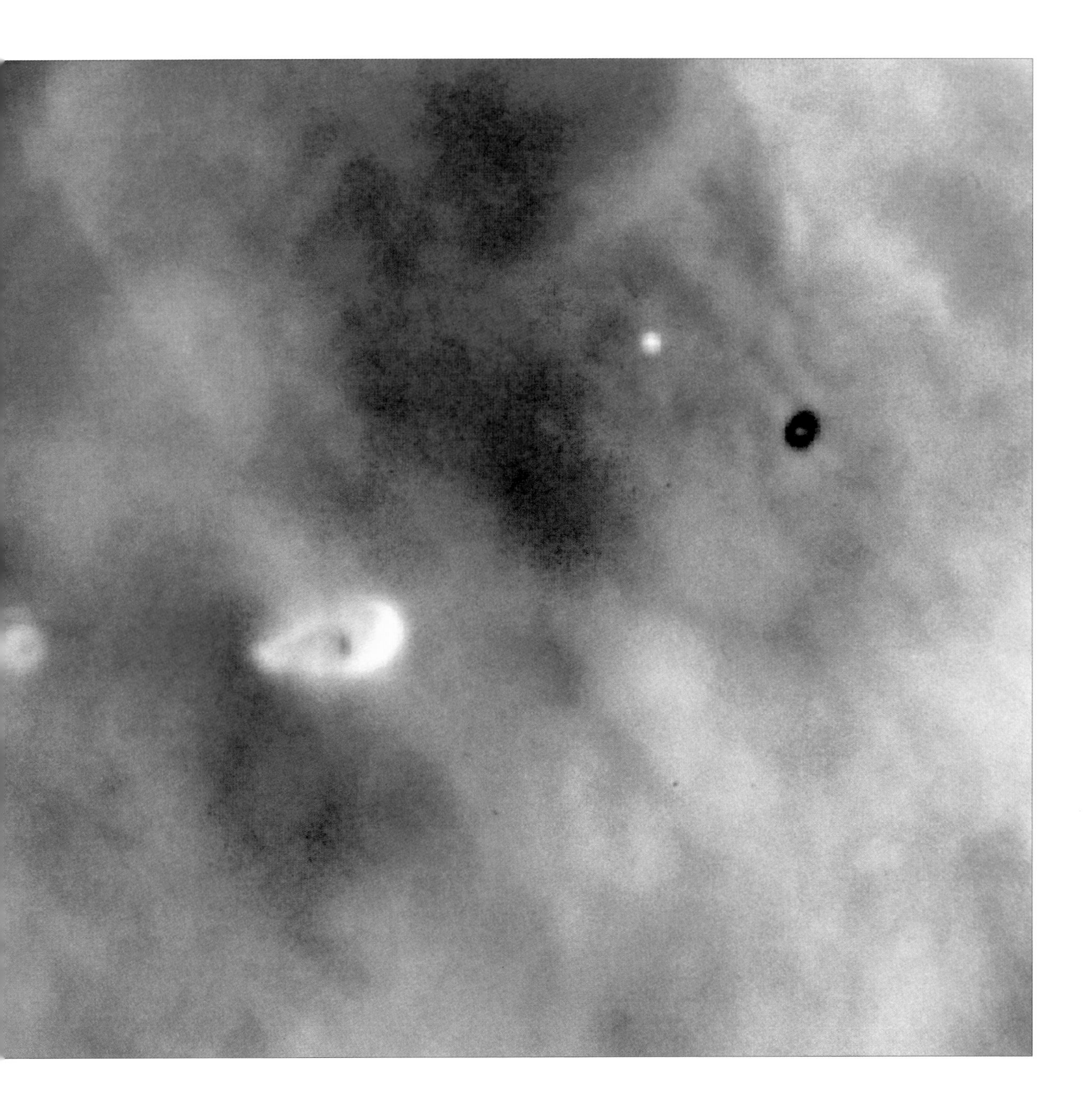

hidden at the centre. Local examples are Jupiter, Saturn, Uranus and Neptune.

The images of the typical proplyd taken from the Orion nebula show that the dust surrounding the new-born star has too much spin to fall into the star itself. Instead, it is spreading out into a broad flattened disk – precisely the configuration needed for planetary formation. In one image, Hubble

Five young stars within the Orion nebula seen close-up. Four are surrounded by protoplanetary disks, which may – if conditions are right – eventually evolve into new systems of planets circling their parent star, much as the planets of our own solar system did approximately 4.5 billion years ago.

Credit: C. Robert O'Dell (Rice University), and NASA

caught an elliptical disk in silhouette against the bright background of the nebula. Dr O'Dell remarked that this particular picture 'represents the most direct evidence uncovered to date for proto-planetary disks'. Using the fine resolution of the space telescope, he was able to measure how much material was present in the disk. The dusty disk itself is almost eight times bigger across than our own solar system, stretching out about 26 billion miles (42 billion kilometres) into space. But the star at the centre of the disk in question weighs only about one-fifth that of our own Sun and is comparatively cool, shining with a reddish glow.

Dr O'Dell found that the young stars that are accompanied by proplyds tend to be about the same size as the Sun or smaller, and he concludes that bigger and hotter stars probably vaporise their own disks before planets can form. Other astronomers have suggested that because the Orion nebula contains numbers of very big and very hot stars, most of the proplyds discovered there are indeed being boiled away (at temperatures reaching more than 10,000 degrees Kelvin*) by their energetic neighbours. At a rough guess, it might take 10 million years to turn out the first planets; according to these theories, most of the proplyds will be done for within a million years. Dr O'Dell's team hotly disputes these suggestions, but everyone accepts that there are several proplyds in the Orion nebula that do not appear to be under lethal attack from nearby big and dangerous stars, and which are therefore likely to go on to produce planets. The argument is about numbers, not about the fundamental theory of planetary formation; everybody agrees that the work done by the Hubble teams has confirmed that the right conditions for forming planets will exist across the universe.

With strong evidence supporting the theory of planetary formation, the next task is to find some fully formed planets beyond our solar system. One of the strongest candidates is, in astronomical terms, just around the corner from Earth.

Beta Pictoris is a star found in a constellation called the Painter's Easel, visible only in southern skies, and is about fifty light-years away, compared to the 1,500 light-years of the Orion nebula. In 1983 the infra-red astronomy satellite IRAS discovered that a broad disk of dust hundreds of billions of miles in diameter is orbiting the star. Thus alerted, astronomers have been training their ground-based optical telescopes on the star ever since, and have verified the existence of a disk (seen edge-on from Earth's perspective) which contains some carbon-based chemicals and which, it was concluded, could be a solar system in the making. They found that there was an inner zone within the disk that had been swept clear of dust and debris, and wondered whether one or more unseen planets might be responsible for clearing up. The problem was that the disk itself is faint and therefore hard to examine; the microscopic grains of ice and dust making up the disk are visible only because they catch and reflect the light of the star.

When the sharp eyes of the Hubble Space Telescope were trained on the disk, everything literally sprang into focus. Images of the disk ten times sharper than had ever been seen before began

* Scientists use the Kelvin temperature scale which differs from Centigrade or Celsius only in that the Kelvin scale starts at absolute zero whereas the Celsius scale had its zero at the freezing point of water — roughly 273 degrees above absolute zero. When measuring temperatures of thousands of degrees, the difference does not really matter.

Opposite: A comparison between images of the star Beta Pictoris obtained by ground-based telescopes (bottom) and Hubble (top) shows the extra clarity obtained by the space-based telescope. Both images show a sideways view of the vast disk of dust circling the star, which may house one or more unseen planets.

Credit: Top, Al Schultz (CSC/STScI) and NASA); Bottom, Paul Kalas (University of Hawaii)

Jupiter, the largest planet in the Solar System, does not have the dramatic ring system so evident in even the smallest telescope's view of Saturn. However the characteristic banding patterns in its upper atmosphere give it its distinctive appearance. These represent separate zones in some of which the winds blow east, whereas in adjacent ones the winds blow in the opposite direction.

Credit: NASA

arriving, and for the first time detailed pictures emerged of variations in the shape of different parts of the disk. In 1996 it was announced that Hubble had uncovered a warp in the inner part of the disk of dust. The best explanation is that the disk has been disturbed by the gravitational pull of a planet. Breaking the news, Dr Chris Burrows of the European Space Agency, stated: 'The presence of the warp is strong, though indirect, evidence for the existence of planets in this system. If Beta Pictoris had a solar system like ours, it would produce a warp like the one we see. The Beta Pictoris system seems to contain at least one planet not too dissimilar from Jupiter in size and orbit.'

It is entirely possible that rocky planets like Earth might also be orbiting the star, although there is no evidence (even indirect) for these as yet. Dr Burrows pointed out that it is almost impossible to view a planet around another star directly; their existence always has to be inferred because 'any planet will be at least a billion times fainter than the star, and presently impossible to view directly, even with Hubble'.

Before Hubble discovered the scores of proplyds around the stars in the Orion nebula, the disk round Beta Pictoris was one of only four that had been definitely confirmed in the Milky Way. IRAS had found evidence of disks round three other stars: Vega (or Alpha Lyrae, the fifth brightest star in the sky); Fomalhaut (Alpha Piscis Austrini); and Epsilon Eridani. Indirect evidence for the existence of fully formed planets round other stars than these has come from many years of observations with radio telescopes and special instruments attached to ground-based optical telescopes. For example, it is believed that three small planet-sized objects are orbiting a distant star known (not very romantically) as PSR 1257+12. This is an exotic type of star known as a pulsar, because it emits radio waves in characteristically regular pulses as it rotates at high speed. (In fact, so regular are pulsars' transmissions that, when they were first discovered, it was thought that these signals might be from extraterrestrial civilisations trying to make radio contact.) The pulsation rate of PSR 1257+12 is slightly irregular and astronomers interpret this as due to the perturbing effect of the gravitational pull of three small planets, one the size of the Moon and two more that are between two and three times the mass of Earth. Decades of work by Russian astronomers looking at another pulsar, PKS B0329+54, suggest that it is being circled by a planet about twice the mass of Earth.

The problem with pulsars is that they are the remnants of old stars that have blown up, and so any original planets circling them will almost certainly have been consumed in the fireball; the planets we think we can detect today are likely to have been formed after the big explosion, and so hold little hope of having developed life as we know it in our own solar system.

In recent years other astronomers have begun looking for planets by searching for tiny shifts in the spectrum of light being produced by stars, assuming that these wobbles are being caused by the pull of one or more planets (a controversial assumption, as others believe that the wobbles are caused by upheavals purely within the bodies of the stars themselves). By the middle of 1997 claims had been made for the discovery of at least eight planets orbiting other stars than our own. But there are puzzling oddities and inconsistencies in the measurements taken by astronomers. The planets around some of the stars appear to be heavier than Jupiter, yet orbiting closer to their stars than Mercury (a tiny planet by comparison) orbits the Sun. Other planets appear to have elongated, oval-shaped orbits around their stars, whereas the planets in our solar system have orbits that are nearly circular.

The arguments rage back and forth. The fundamental problem is that the only system of planets-plus-star that we can (so far) observe properly is our own. A sample of one is hardly a basis for conclusive analysis. If we find indirect evidence for planets around other stars that does not fit in with our own local situation, should we conclude that the 'evidence' is in fact wrong and there are no

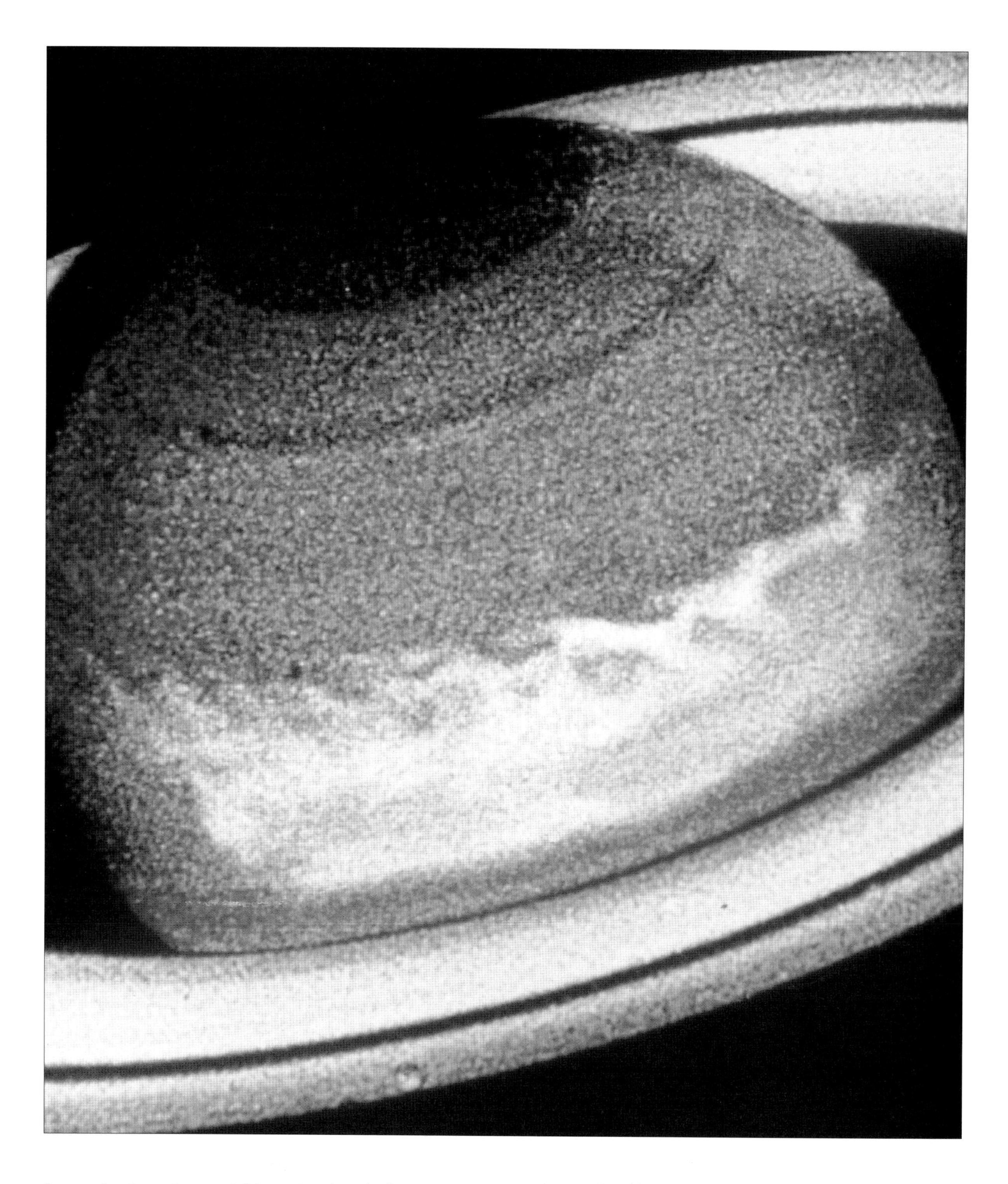

Storm clouds on Saturn visible as the white fluffy disturbances not far from the equatorial region. Hubble can make regular observations of our planetary neighbours, tracking changes almost as they happen. This is a valuable extension to the knowledge gleaned by visiting space probes, such as Voyager and Cassini. Their observations have the advantage of being taken from close up, but they tend to be made over shorter periods of time.

Credit: NASA

planets there, or that our solar system may not be typical?

Some astronomers have suggested that our solar system may be exceptional in being so stable, that it may be both middle-aged and boring. (Boring only in astronomical terms, we hasten to add; an astronomer's idea of 'exciting' usually involves incredible violence and cataclysmic fireworks visible from the other side of the universe. From a life-form's point of view, boring is brilliant.) These scientists speculate that it may not be normal for planets to occupy their own distinct, stable orbits, but instead to career around their solar systems smashing into each other like some crazy and lethal game of cosmic billiards.

If they are right, then the prospects for life on planets beyond Earth, while still entirely possible, are less bright. Had Jupiter, for example, had an oval orbit instead of a circular one, then it could have collided with Earth by now, and had life formed by then on Earth, that would have been the day that it all came to a very sudden end.

Even so, there is strong evidence that perhaps our solar system is not nearly boring enough for comfort. It is now widely accepted that the most successful dominant form of life on Earth to date – the dinosaur – was wiped out some 65 million years ago by a single asteroid just over 6 miles (10 kilometres) across which crashed into the Earth. This impact threw up so much dust and smoke that it immediately changed the climate, blotting out the Sun's light and shrouding the planet in years of winter. Many plants died; so the animals that ate the plants starved; so the animals that ate the animals that ate the plants starved too. In short order, the planet was instantly rendered uninhabitable for dinosaurs.

Of course, what was terminally bad news for the dinosaurs was very good news for us; with the competition removed, some small, rat-like mammals that had survived the meteorological mayhem were able to prosper and evolve rapidly into the myriad forms we see today – mice, whales, humans – and take over the planet. But there is nothing to stop another meteorite of comparable size hitting the Earth and wiping out mammals, including humans, and so allowing (say) insects to inherit the Earth.

We can take some comfort from the fact that such impacts occur here only about every 100 million years. However, even small fragments of matter can cause terrific damage. For example, in June 1908 a forested region of Siberia 25 miles (40 kilometres) across was devastated by an enormous explosion that resembled nothing so much as a nuclear bomb blast, with rank after rank of trees being felled in lines radiating away from the impact point. It is now believed that the cause was a fragment of the tail of Encke's comet which had broken off and fallen to Earth.

How likely is it that we might be subjected to a repeat performance? The concentration of the world's scientists was focused sharply on to this question by the spectacular impact of comet Shoemaker-Levy 9 on a fellow planet – Jupiter – in 1994, an event that was tracked by the Hubble Space Telescope.

The Sandia National Laboratories in Albuquerque, New Mexico, is part of the United States' nuclear weapons research and manufacturing complex. A Sandia scientist, Dr David Crawford, has been asking the world's fastest computer – an Intel Teraflops machine – what would happen if Earth was hit by a single cometary fragment a kilometre (1,000 yards) across, i.e. the size of just one of the many fragments that hit Jupiter in 1994. (The Teraflops machine is the only one capable of working out the effects properly: Teraflops means that it can add one million million numbers together each second.)

Dr Crawford's calculations showed that such a fragment landing in the sea would impact with all the force of 300,000,000,000 tons of TNT, about 100

Hubble observes the heart of the comet Hyakutake, as it swung within 15 million kilometres (9.3 million miles) of Earth. The main frame shows the cometís dusty tail sweeping up to the top-left-hand corner, where three fragments have broken off and created their own tails. Top right shows a close-up of fragments, and bottom right is a close-up of the nucleus, showing an area 760 kilometres (470 miles) across.

Credit: H. A. Weaver (Applied Research Corporation), HST Comet Hyakutake Observing Team, and NASA

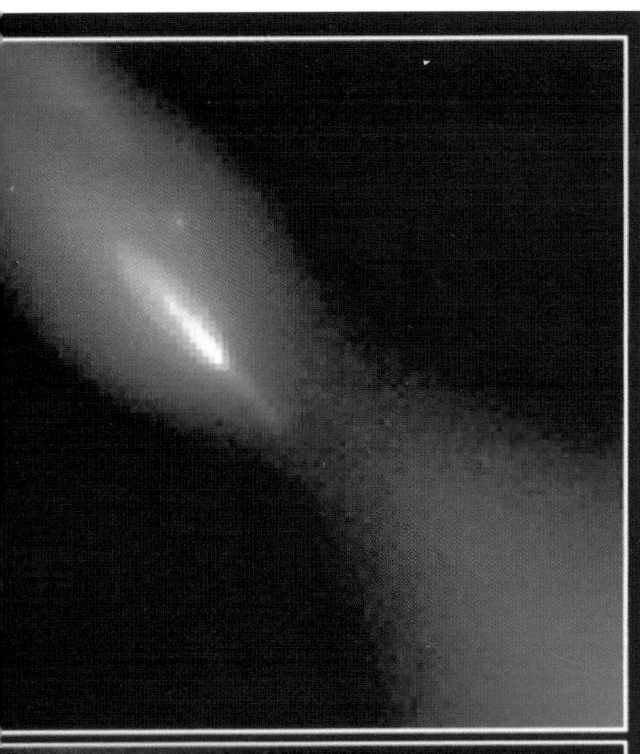

times the entire explosive power of all the nuclear weapons in existence at the height of the Cold War. The impact would vaporise the comet (a billion tons of rock and ice) together with 300 to 500 cubic kilometres of ocean which would flash to steam in a huge explosion. Dust and water vapour would be sent round the earth; some would even bounce back out into space. It would leave a dent in the floor of the ocean.

The calculations indicate that a kilometre-sized chunk of comet is just slightly too small to trigger a global winter and mass extinctions of plant and animal life. However, humanity might not escape quite so easily if the cometary fragment came down on land and started widespread smoky fires and scattered dust.

While dinosaur-killing asteroids 10 kilometres wide are once-in-100-million-year events, the solar system is littered with loose fragments a kilometre wide. So much so, indeed, that one is likely to hit the Earth every 300,000 years. Or, to put it slightly less comfortingly, there is a 1 in 3,000 chance that we will be hit by such a fragment during either our or our children's lifetimes.

Part of the Hubble Space Telescope's work has been to track down the lairs from which such life-destroying comets might come. In 1995, out beyond the orbit of Neptune (then the most distant planet from the Sun), Hubble found evidence for a primordial reservoir of comets, possibly as many as 200 million of them, essentially unchanged since the birth of the solar system more than 4 billion years ago.

More than forty years ago, the Dutch-born astronomer Gerard Kuiper (who spent most of his productive scientific life in the United States) suggested that there might be a band of comets lurking at the outermost fringes of the solar system. But the existence of the Kuiper Belt remained an imaginative hypothesis until 1992 when (ground-based)

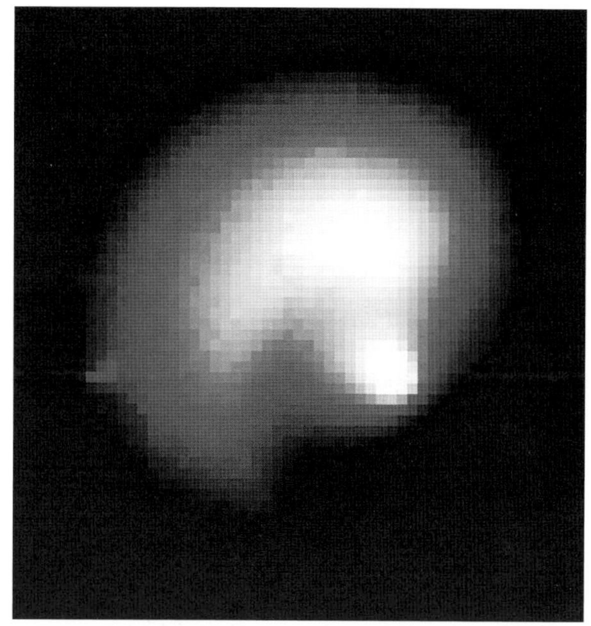

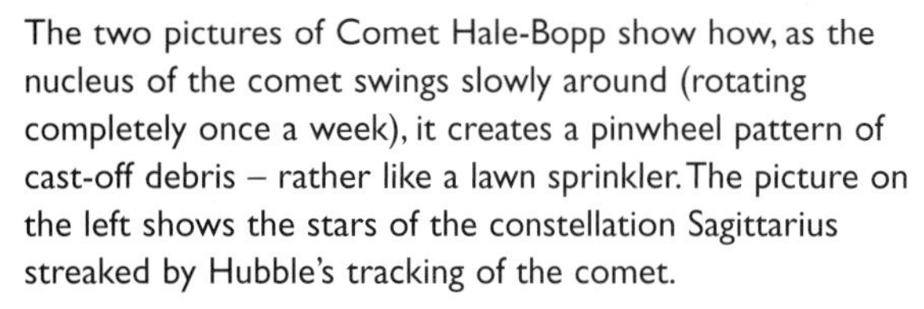

The two pictures of Comet Hale-Bopp show how, as the nucleus of the comet swings slowly around (rotating completely once a week), it creates a pinwheel pattern of cast-off debris – rather like a lawn sprinkler. The picture on the left shows the stars of the constellation Sagittarius streaked by Hubble's tracking of the comet.

On the right, the close-up shows a bright clump of debris outshining the nucleus below it. The debris is probably an icy chunk of the comet's crust which has disintegrated into a cloud of bright particles.

Credit: H.A.Weaver (Applied Research Corporation), P.D. Feldman (Johns Hopkins University), and NASA

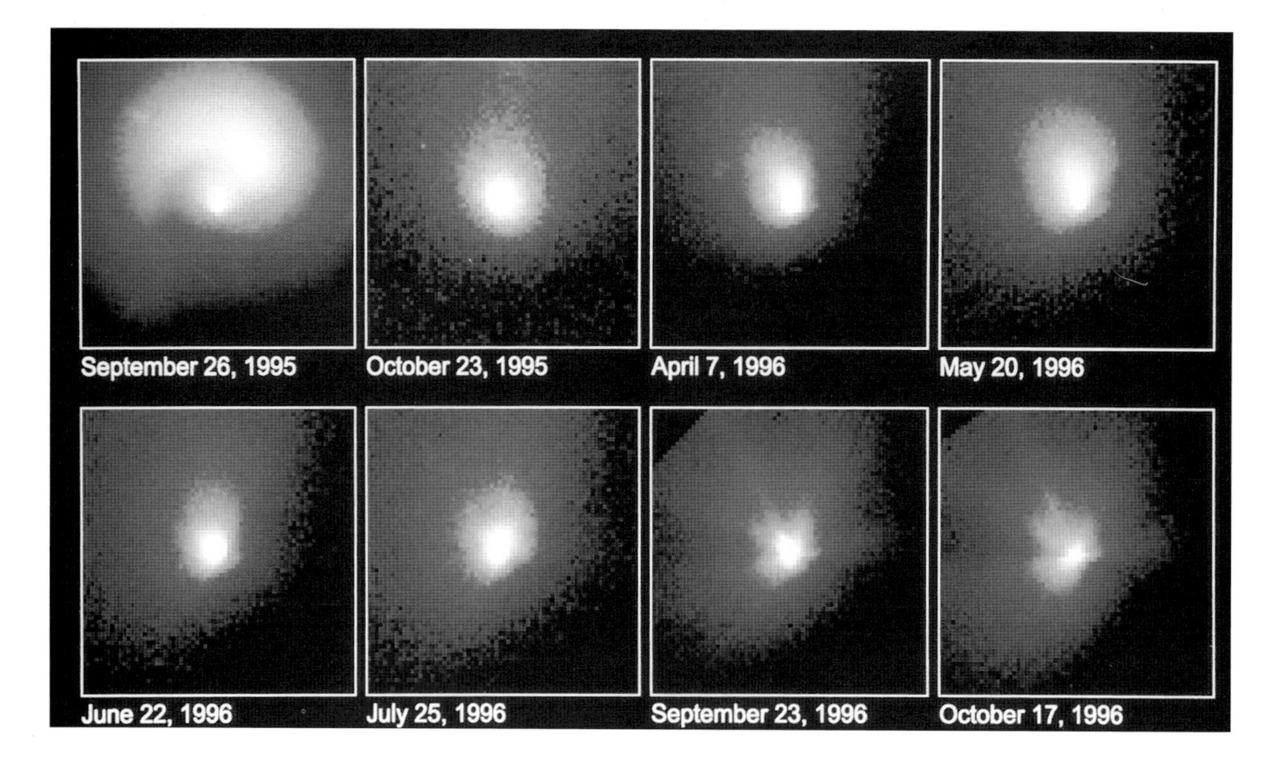

telescopes picked up evidence for large icy objects as much as 200 miles (320 kilometres) across. However, only Hubble's sensitivity and ability to resolve small objects gave astronomers the opportunity of searching for the smaller bodies that they knew had to be there, and which they finally located in 1995.

The astronomers engaged on the comet hunt are, in effect, looking for the remains of our very own protoplanetary disk, the one from which our solar system was created. The nuclei of the comets are made of the primordial stuff – dust, ice, gas – which in other parts of the solar system coalesced to become full planets. One of Hubble's comet hunters, Dr Hal Levison of the Southwest Research Institute in Boulder, Colorado, put it like this: 'The Kuiper Belt is the best laboratory in the solar system for studying how planets are formed. We believe we are seeing a region of the solar system where the accumulation of planets fizzled out.'

The comets of the Kuiper Belt are far from the Sun. The region stretches from just beyond the orbit of Neptune to some 500 times the 95 million miles (150 million kilometres) that separates the Earth from the Sun. So little of the Sun's heat penetrates that the comets are very cold indeed. When a comet swings closer in to the centre of the solar system, the Sun's increasing warmth boils off material from the icy head. Like dry ice on Earth, the material moves straight from solid to gas without passing through a liquid stage. As the Sun's rays liberate dust and gas from the nucleus of the comet, it appears brilliant in the sky with a bright head and long tail of matter trailing behind it.

Opposite: Hubble images of Comet Hale-Bopp as it evolved over two years, with changes being prompted by the increasing warmth as it sped closer to the Sun. The pictures taken in the autumn of 1996 show multiple jets of gas and debris coming from the nucleus, probably caused by the opening up of multiple vents on the surface of the solid nucleus (which, being only 40 kilometres or 25 miles across, is not actually visible in these pictures).

Credit: Harold Weaver (Johns Hopkins University), and NASA

As the invading comets come sweeping in from the frozen outer edges of our system, Hubble has been examining them for details about their origin (and, hence, possibly the origins of the solar system itself), and about their behaviour under the stress of solar radiation. When comet Hale-Bopp – one of the brightest comets to approach the Earth in recent years – was still 600 million miles (960 million kilometres) distant, out beyond the orbit of Jupiter, Hubble was able to take pictures of pieces breaking off the central nucleus and dissolving in a shower of 'sparks' – a bright cloud of particles. It also found that the solid nucleus was rotating at a rate of one revolution a week. The telescope was also able to picture comet Hyakutake when it closed with the Earth, showing jets of gas puffing out from the core and pieces breaking off under the stresses.

And Hubble gave us a ringside seat when the most famous comet of recent years, Shoemaker-Levy 9, crashed into Jupiter. The impact sites themselves were hidden just round the back of the planet, but Hubble was able to follow all twenty-one fragments of the comet until a few hours before impact, showing that they remained intact even under the stresses of Jupiter's tremendous gravity until they hurtled into its atmosphere. The analysis of the impacts, which were spread over several days, cast new light on Jupiter itself, in that Hubble was able to use the marks left by the impacts – which persisted for more than a month – to trace the movement of winds and atmosphere. But one episode also revealed much about the nature of Shoemaker-Levy 9 itself.

Four days before impact Hubble detected a mysterious flash of light from one cometary fragment – nucleus G – when it penetrated the outer reaches of Jupiter's powerful magnetic field, still some 2.3 million miles (3.7 million kilometres) above the planet's surface. By studying the spectrum of the light emitted during the flash, astronomers were able to pick out the characteristic signatures of different chemical elements. They found some things that they expected, like magnesium ions. But they also found that one thing they had expected to see was mysteriously absent: water. This despite the fact that comets are thought to be largely made up of water ice. And when Hubble analysed the elements visible at the impact sites it found silicon and iron, elements not normally present in Jupiter's atmosphere in any detectable quantities.

Although astronomers still favour the idea that Shoemaker-Levy 9 was indeed a comet, Hubble has raised the tantalising suspicion that it just could have been a rogue asteroid.

Asteroids are the second main class of fragments in our solar system that failed to make the grade as planets. Comets are largely icy because they originate so far from the Sun, whereas asteroids are virtually ice-free because they form so close to the Sun that any water has long ago been driven off by its heat and radiation. For a long time people had assumed that the asteroid belt located between Mars and Jupiter was the debris of a planet that exploded long ago. Today, astronomers believe instead that the asteroids represent the building blocks of a planet that never formed, probably due to the disruptive gravitational pull of Jupiter.

Although the total amount of material present in the asteroid belt is only one-twentieth of the Earth's own Moon, there are thought to be about a million asteroids bigger than 1 kilometre across. Ceres, the largest asteroid, is a whopping 630 miles (960 kilometres) in diameter, and was discovered by the Italian astronomer Guiseppe Piazzi in 1801. Apart from their lack of size, asteroids have many of the characteristics of 'proper' planets. They are rocky, and some appear to have undergone some heating after their formation, with the heat probably coming from the decay of a radioactive form of aluminium. In addition, many of them bear the scars of violent encounters with other asteroids and flying fragments.

In 1995 Hubble was able to map the face of Vesta, brightest among the asteroids. Although Vesta is only 320 miles (515 kilometres) in diameter, the evidence of internal heating is plain to see. Hubble found traces of lava flows, confirming that the interior of Vesta was once molten, and pushed out lava on to the surface where it cooled and is visible today. In some places the outer crust has been stripped away by the repeated impact of collisions with meteors and other asteroids to reveal the mini-planet's interior – the most ancient geological formations visible anywhere in the solar system. Not only that, but the scientists analysing the Hubble images of Vesta have the great advantage of being able to refer back to fragments held in the laboratory. One of those collisions ripped chunks out of the asteroid's surface, and sent them flying down in a fiery ball into a field in Australia in 1960. The pieces were recovered ten years later, allowing us to work on rocks from four bodies of our system: Earth, her Moon, Mars, and Vesta the asteroid.

Hubble's observations of the asteroids could be the starting point for some very practical applications rather than remaining for ever 'pure science'. In late 1997 an American corporation, SpaceDev, announced that its business plan included an intention to stake a commercial claim to some asteroids. The company had been set up by an American developer of computer software, James Benson, the founder of Compusearch Software, which specialised in full text indexing and search engines (predecessors to Internet web search engines). Mr

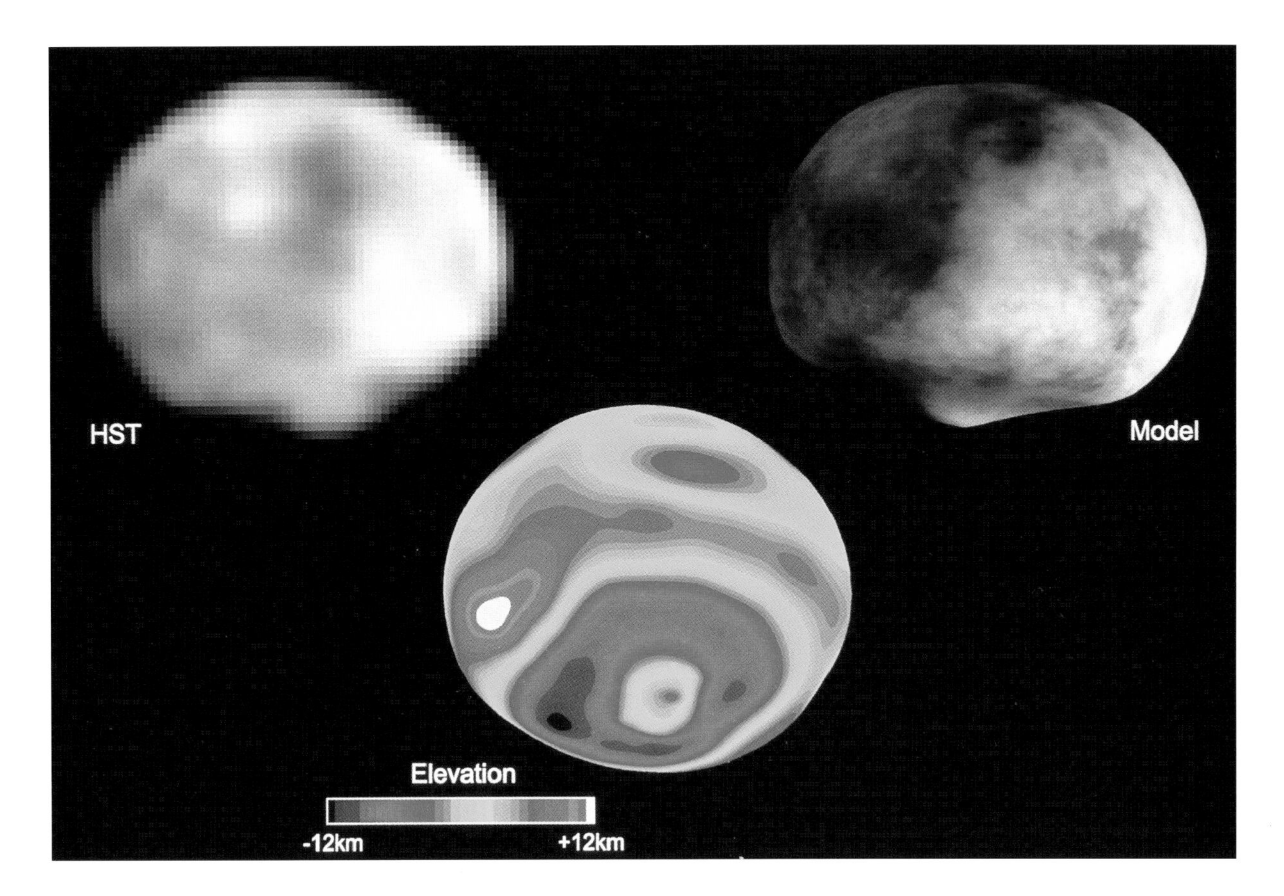

The asteroid Vesta, showing an enormous crater almost 8 miles (13 kilometres) deep created by an ancient impact. Had the Earth suffered an impact on a similar scale, the crater would have spread right across the entire Pacific Basin. The topographical map (bottom) shows a mini-peak in the centre of the crater, created when molten rock slopped back from the rim after the impact and cooled.

Credit: P. Thomas (Cornell University), B. Zellner (Georgia Southern University), and NASA

Benson's new company intends to send a small spaceprobe, the Near Earth Asteroid Prospector (NEAP), to the asteroids. If all goes according to plan, NEAP will be the first private spacecraft ever to leave Earth's orbit, the first to visit another planetary body, and the first private spacecraft to land on another planetary body. SpaceDev estimates that the entire mission will cost less than $50 million and that it can reap a profit from the sale of the scientific data that the probe will send back.

At least to begin with, SpaceDev is not intending to start mining and extracting valuable minerals from the asteroids, but the company made clear that this is the long-term aim. At SpaceDev's inauguration, a company spokesman stressed that 'an objective of NEAP is to focus international corporate attention on the enormous value of near earth asteroids. As energy, mining and natural resource companies begin to incorporate future space resource utilisation into their corporate planning horizons, SpaceDev will be positioned to perform inexpensive but profitable private resource surveys and assessments for such companies, much as oil companies pay for private geologic and seismic surveys and assessments.' It is not clear how realistic SpaceDev's plans are at the present state of technology, but its formation is a clear indicator that space may one day – possibly quite soon – become a commercialised region.

This little chunk of rock is thought to be a fragment of crust blasted from the asteroid Vesta, and sent flying through space until it was eventually tugged by Jupiter's gravitational pull into a trajectory that took it to Earth, where it landed as a meteorite in a field in Australia nearly 40 years ago. Its importance cannot be overstated; scientists have access to laboratory samples of only three other large bodies from our solar system – Mars, the Moon, and of course Earth.

Credit: R. Kempton (New England Meteoritical Services)

Which raises a number of further questions. Will men and women travel in the same way to the planets, not as visiting scientists but as workers or colonists, to carve out fortunes or build new populations beyond Earth? Do the other planets in our own back yard offer any future for us apart from being fascinating laboratory specimens? What can we learn about the nature of other planets circling other stars by looking at our planetary neighbours? And will that knowledge help us to track down planets where life is more rather than less likely to have occurred?

The first thing to say about our planets is that, from the point of view of any earthbound life-form, they are extremely unfriendly. A classic way to remember the order of the planets is the (admittedly slightly daft) expression Mother Very Earnestly Made a Jam Sandwich Under No Protest, which gives you the running order (from the closest to the Sun to the most distant) Mercury, Venus, Earth, Mars, Jupiter, Saturn, Uranus, Neptune, Pluto. (Confusingly, Pluto's orbit means that for the past twenty years it has been a tad closer to the Sun than Neptune.) But whichever way you list them, they are all nasty places to find yourself.

The life expectancy of an unprotected human stepping on to the surface of our neighbouring planets would vary between a second and maybe a minute, depending on whether one is to be frazzled to a crisp, squashed flat as a sheet of paper, dissolved in acid, boiled, asphyxiated, irradiated, turned to a block of ice, or an interesting combination of several of these unpleasant occurrences. In a state-of-the-art spacesuit, and given a suitable supply of oxygen and nourishment, the odds are slightly improved; on Mars, our intrepid explorer might be able to live indefinitely, provided he or she could dodge the storms and avoid the hard radiation coming in from space. On any of the other planets, the best that one can say is that our suited-up explorer might, sadly, live just long enough to appreciate the full folly of making the voyage.

Take Venus, for example. Hubble has taken a hauntingly beautiful image of a crescent Venus, the brilliant clouds reflecting the Sun's light from the upper regions of the atmosphere. Many hopeful science fiction writers have imagined that there might be life on Venus, which they have tended to portray as being covered in steamy but lush jungles. However, chemical analysis of the upper atmosphere – undertaken by the Goddard High Resolution Spectrograph on board Hubble – show that sulphur dioxide ejected from Venus's many volcanoes is being broken up by the Sun's light and then washed back to the surface as acid rain, in the same process as happens on Earth. This particular rain is concentrated sulphuric acid and, as if that were not enough, any traveller would have to cope with immense pressures – equivalent to being 3,280 feet (1,000 metres) under the sea on Earth – and temperatures of around 465 degrees Celsius. It would also be dark, for only one-fiftieth of the sunlight shining on the top of the clouds filters down to the surface.

The Soviet Venera probes during the 1970s actually made it through the dense cloud layers to the surface, relaying TV pictures back home and scooping up samples of the Venusian soil for brief chemical analysis. The evidence is all too depressing; despite its apparent beauty, Venus is positively hellish. So too is Mercury, which is so close to the Sun that a block of lead dropped on to the surface would turn to liquid and flow like water (real water, of course, would be not so much evaporated as annihilated into its constituent molecules in a flash). On the shady side of the planet, in contrast, the temperature is worse than freezing: around minus 180 degrees Celsius.

And what of the four gas giants: Jupiter, Saturn, Uranus and Neptune? The truth is that, apart from Jupiter (which as well as being the biggest planet is the closest of the four to observers on Earth), remarkably little detail has been revealed about these massive bodies until recent years, when, among others, Hubble has played its part in telling us more about what is going on in our own back yard.

When Galileo first looked through his telescope at Saturn in 1610, his crude apparatus showed him a triple image which he took to be three companion planets right next to each other, but were actually the image of the planet plus its famous rings visible on either side. The presence of the rings was correctly analysed in 1655 by the Dutch astronomer Christian Huygens, and they have fascinated and baffled astronomers ever since. Are they solid? Are they in one piece or several rings with gaps between them? How do they stay in place under the gravitational pull of the second largest planet in the solar system (ninety-five times as heavy as Earth), to say nothing of its twenty moons?

Today astronomers know that all the gas giants have their own ring systems, although none is as spectacular as Saturn's. The rings are actually a huge expanse of crushed ice, ranging in size from ice-cubes to lumps as big as refrigerators. They start about 4,000 miles (6,500 kilometres) above the

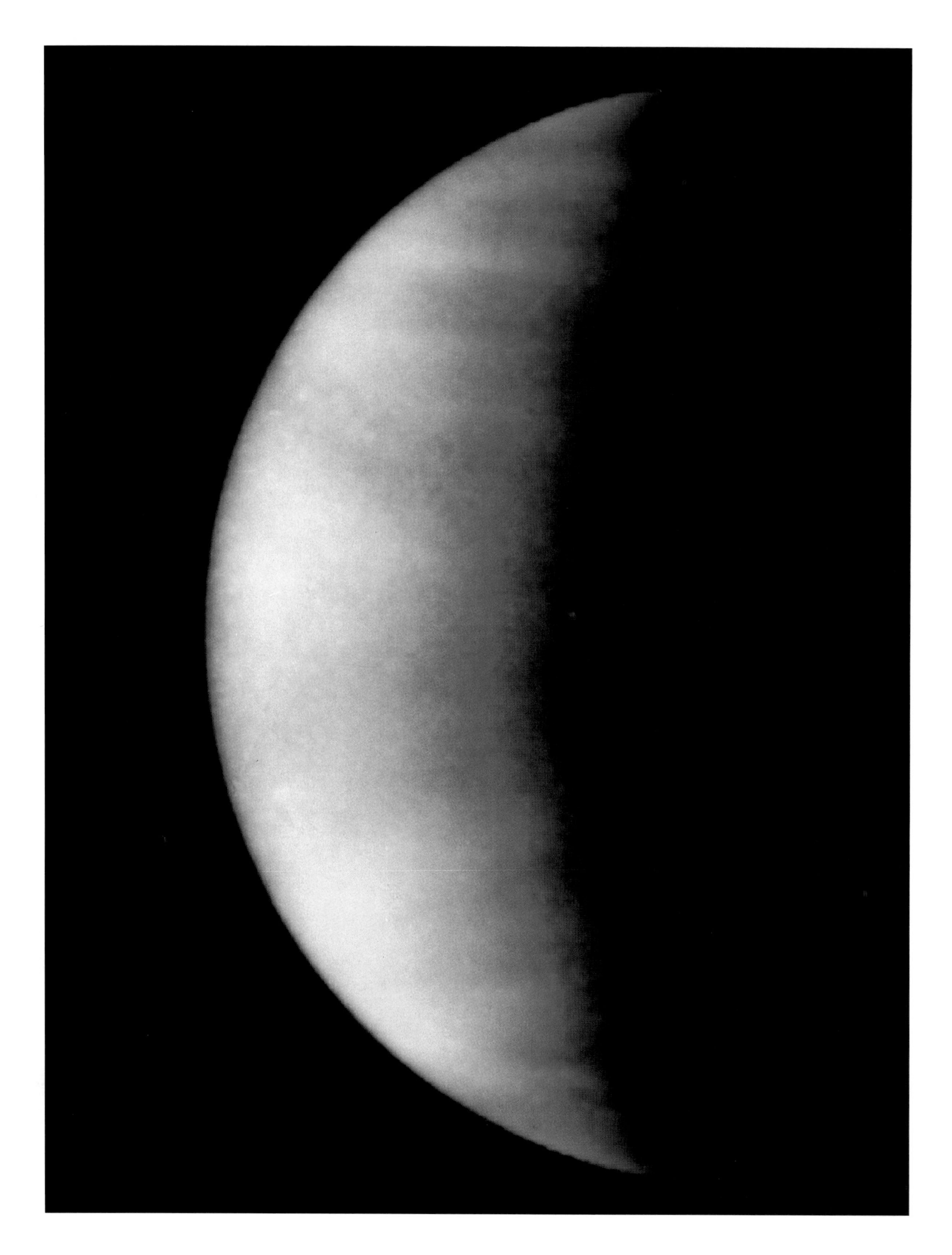

Opposite: Venus in ultraviolet looks stunning – but the vision of beauty hardly reflects the hellish conditions on the surface, which is drenched by acid rain and scorchingly hot. The planet is entirely shrouded by clouds – and these clouds are made not of water vapour, as on Earth, but sulphuric acid.

Credit: L. Esposito (University of Colorado, Boulder), and NASA

Above: Hubble clearly picks out a rare storm on the planet Saturn. The storm, which shows up as the bright area, is about 7,900 miles (12,600 kilometres) across – about the same diameter as Earth. The storm's western edge is being penetrated and distorted by Saturn's strong winds.

Credit: Reta Beebe (New Mexico State University), D. Gilmore, L. Bergeron (STScI), and NASA

tops of Saturn's clouds (which are mostly made up of ammonia crystals) and extend some 46,000 miles (74,000 kilometres) out into space, although each ring is extremely thin. Each of the seven or so main rings is subdivided into ringlets, of which there may be more than 10,000. The millions of pieces of ice girdling the planet are likely to be the remains of a moon that never quite made it, inhibited from forming because it was so close to the planet, and therefore broken up by Saturn's gravitational pull.

When the Voyager space probe hurtled past Saturn almost twenty years ago, even its fleeting glimpses revealed a wealth of new facts, as well as a number of previously undetected moons. Astronomers are rubbing their hands at the prospect of being able to use Hubble to make the first really consistent and close-up examination of the planet and its strange, icy companions.

The same can be said for Neptune, currently the planet furthest away from the Sun. Again, Voyager offered tantalising glimpses of what had been until then a tiny blue disc in the viewfinders of ground-based telescopes; and again, scientists are looking to Hubble to provide the first systematic and steady monitoring of the planet.

Voyager told us that Neptune is a planet of storms, clouds and hurricanes. As it flew past

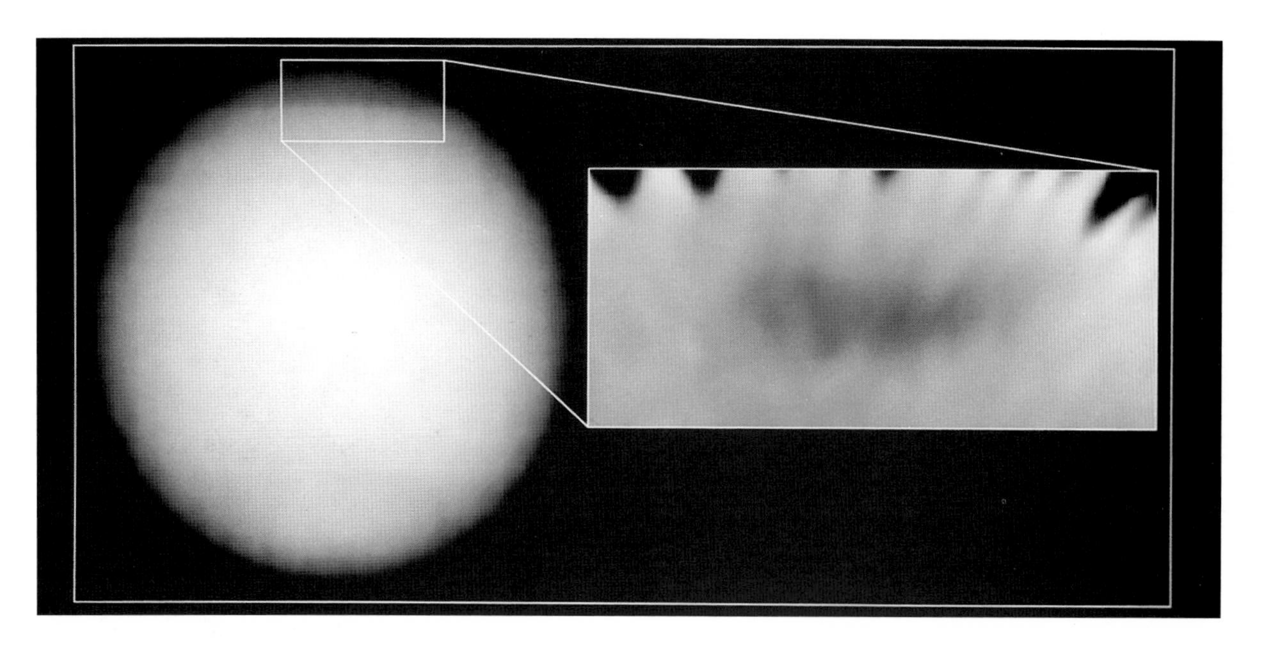

Hubble tracked down a new Great Dark Spot located in Neptune's northern hemisphere, just as it discovered that a similar spot in the southern hemisphere had disappeared in 1994. Nobody knows how long these features are likely to persist. The new spot has clouds all around its edge, formed by gases that have been propelled to high altitudes where they cool sufficiently to form clouds of methane ice crystals.

Credit: H. Hammel (Massachusetts Institute of Technology), and NASA

Neptune in 1989, Voyager detected furious winds gusting up to 730 miles (1,170 kilometres) an hour, bright clouds high in the hydrogen and helium atmosphere, and also two large dark spots which planetary scientists believe are giant hurricanes. The largest of these, in the southern hemisphere, was nicknamed the Great Dark Spot because it was thought to resemble the Great Red Spot on Jupiter, a storm system which has persisted for hundreds of years. But the longevity of Neptune's storm systems was impossible to measure because Voyager took only a brief look, and the ground-based telescopes could not pick out the spot at all because of the blurring effect of the Earth's atmosphere.

When Hubble was first trained on Neptune in 1994, scientists were keen to see what had happened to the Dark Spot over the past five years. To their amazement, it had vanished. A second look revealed that a new Dark Spot had emerged in the northern hemisphere, virtually a mirror image of the situation detected by Voyager. It is as if the atmosphere had been turned upside down.

The dark spots are thought to be something like hurricanes. As with the eye of a hurricane, the spot has a hole in the centre, opening a window into the deeper part of the atmosphere. The bright clouds are at very high altitudes, like thunderheads that have bubbled up from deeper down. But the intense dynamics of the gas giant's atmosphere has raised baffling new questions for planetary scientists. The heat of the Sun is the main driver of weather here on Earth, but this cannot be important on Neptune because it is so far from the warming rays (and also because the planet itself radiates out into space about twice as much energy as it receives from sunlight). They therefore conclude that there has to be a strong internal source of heat on Neptune, and think that this is probably the effect of the planet contracting under the force of its own gravity, which leads to movement in the interior, which

in turn generates heat. Neptune's atmosphere is thought to be so lively because the cloud tops are warmed from within in this way; if the difference in temperature between the top and bottom layers of atmosphere changes even slightly, then rapid large-scale changes in circulation patterns in the atmosphere will follow.

The Space Telescope Imaging Spectograph (STIS) has allowed Hubble to take ultraviolet images with ten times the sensitivity of earlier cameras. Here Jupiter's aurorae are captured as they spiral towards the gas giant's north and south magnetic poles. The glow comes from the interaction between electrically charged particles and the molecules of the planet's upper atmosphere.

Credit: John Clarke (University of Michigan), and NASA

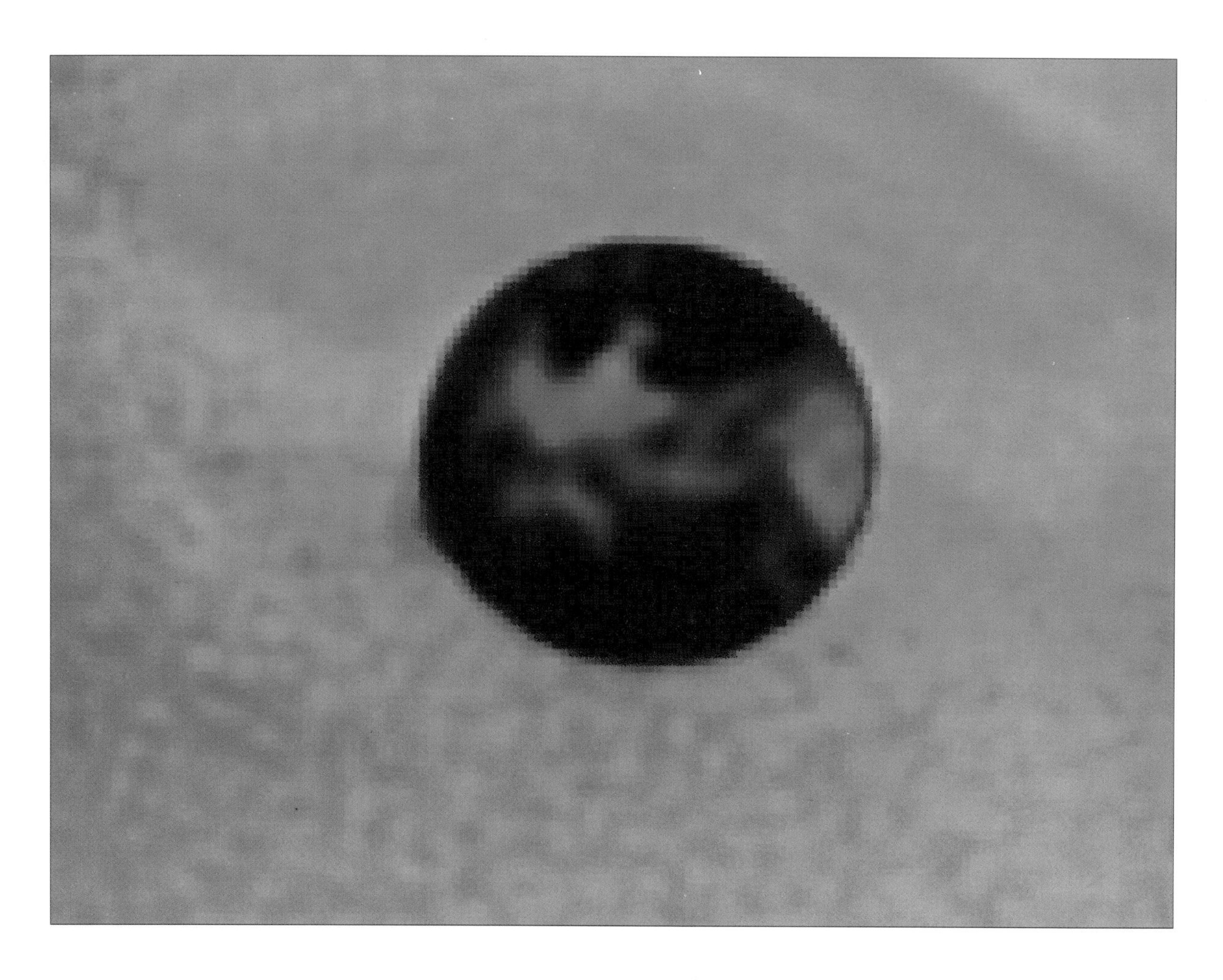

Hubble catches sight of a vast volcanic explosion on Io, Jupiter's large innermost moon. The plume of gas and dust expelled at 3,200 kilometres per hour (2,000 miles per hour) from a volcano called Pele is visible to the bottom left of the moon, silhouetted against the blue background of Jupiter's clouds. The plume is about 400 kilometres (250 miles) high, and is able to reach this height due to the lack of resistance from the thin atmosphere, and the low gravity (which is one-sixth that of Earth).

Credit: John Spencer (Lowell Observatory), and NASA

For many of the planets, one of Hubble's main functions is turning out to be a weather station. Hubble's weather reports for Mars have had a direct and immediately practical application for space missions. For the gas giants, however, of which we know so little, understanding their 'weather' will be the first step towards understanding how they function as planets, and what they can teach us about the likely composition and behaviour of other gas giant planets believed to be circling other stars.

Jupiter is the mightiest planet of the solar system, and also has the most spectacular 'weather'. It weighs 318 times as much as the Earth. All the gas giants have their own original gaseous atmosphere from the time they were created more than 4 billion years ago. (Earth and the other rocky planets, whose original atmosphere was boiled off by the Sun, have 'grown' their own atmospheres since, using gas seeping out from within their rocky cores.) These gas giant atmospheres tend to be composed of lighter elements than those around their rocky companions – hydrogen and helium, the gases found in stars, plus

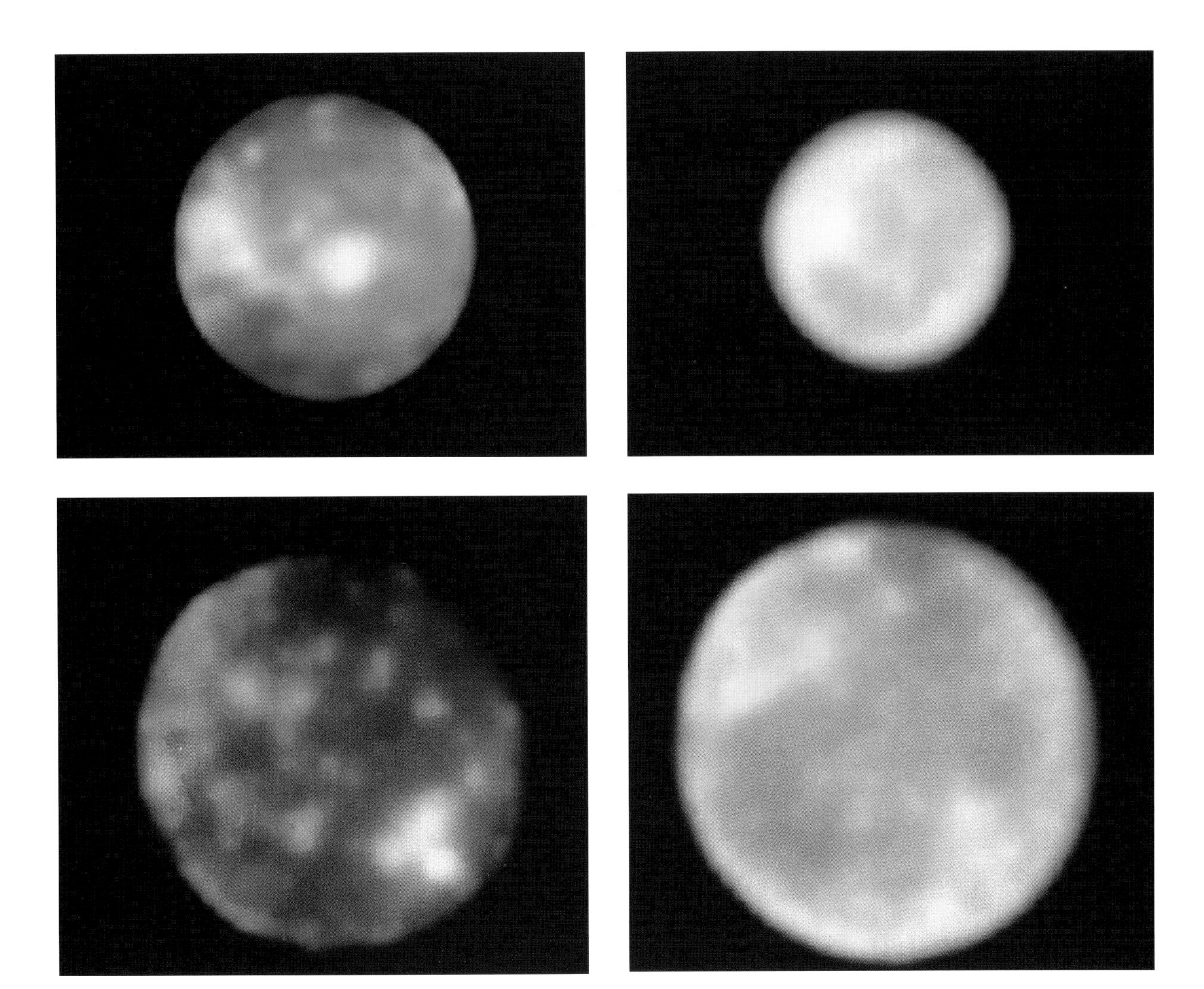

ammonia and methane – but so great are the pressures that much of the gas around the hidden, solid core of each of the planets is compressed into liquid or even near-solid form. Jupiter's sheer size means that the pressure at the rocky surface is 45 million times the atmospheric pressure on Earth. At these pressures, hydrogen assumes a nearly solid metallic state, and the core is surrounded by a huge layer of metallic hydrogen 18,000 miles (29,000 kilometres) thick. Enormous electrical currents surge through this shell, generating around the planet a vast magnetic field; so vast, in fact, that were this 'magnetosphere' visible from Earth, it would fill the same space in the night sky as the full Moon.

Io is one of four 'Galilean' moons first spotted by the great Italian scientist Galileo Galilei nearly 400 years ago. Hubble has found a (very thin) oxygen atmosphere on the surface of Europa, and ozone on the surface of Ganymede. Observations of Callisto show the presence of fresh ice, which may indicate impacts from micrometeorites and the influence of charged particles from Jupiter's magnetosphere. Ground-based telescopes cannot resolve to this detail these four moons.

Credit: K. Noll (STScI), J. Spencer (Lowell Observatory), and NASA

Earthbound telescopes have easily observed the distinctive sets of bands in the upper Jovian atmosphere. In some of the bands, winds blow east, whereas in others the wind is in the opposite

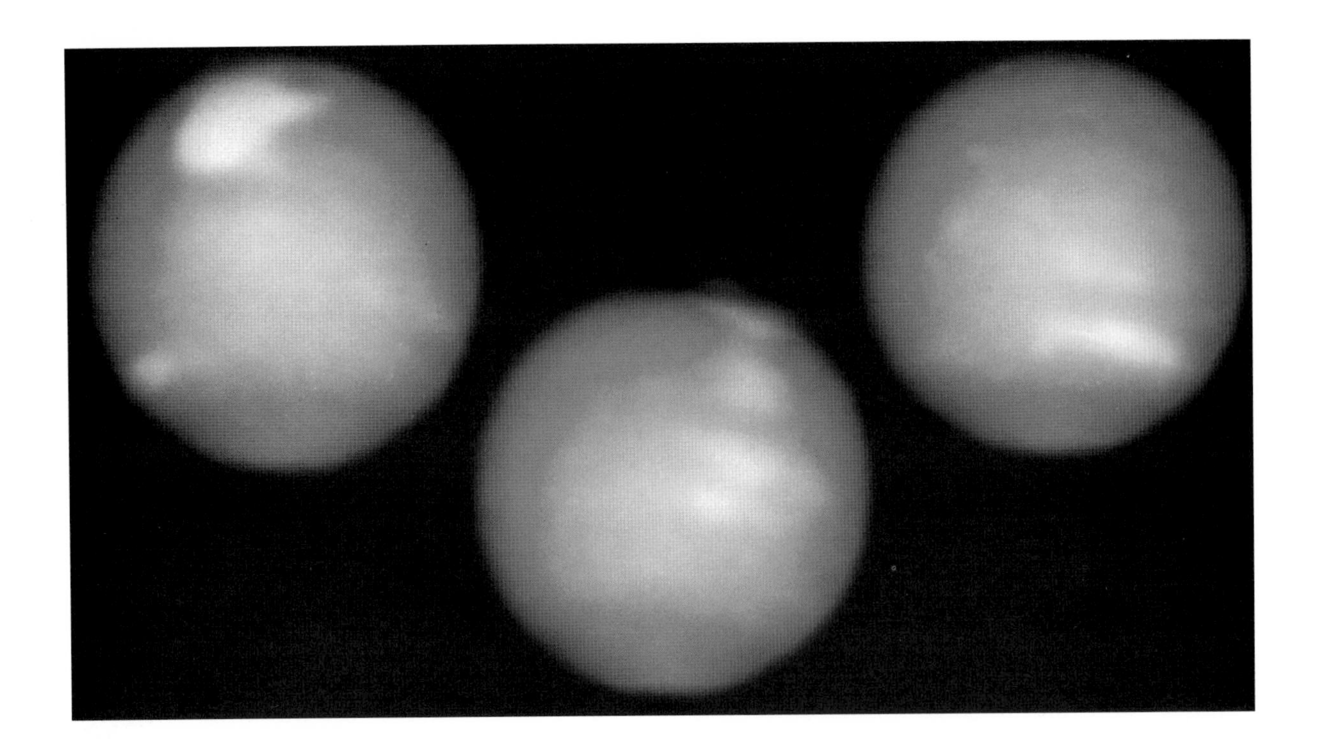

These three images of Neptune were taken over a period of three weeks, and show how the planet's weather is capable of changing over a matter of days. The engine for this meteorological dynamism is probably the stress between the internal heat sources deep within the planet, and the extreme cold of the upper cloud regions. Neptune, the most distant of the gas planets in our solar system, is four times the diameter of Earth.

Credit: J. Hester (Massachusetts Institute of Technology), and NASA

direction. Where they meet, vortices and eddies are visible. The most famous is the Great Red Spot, a swirling storm in the southern hemisphere which was discovered by the English astronomer Robert Hooke in 1664 and which appears to have been raging unabated for the 300 years since then.

The combination of raging winds and vastly powerful magnetic pull must make Jupiter an incredibly violent place. The Hubble Space Telescope cannot 'see' the magnetic field, but it can measure its effects. The telescope's ability to take sharp ultraviolet pictures of dancing luminous clouds of gas high in Jupiter's atmosphere – the 'aurora', produced when magnetically charged particles impact the gas molecules of the atmosphere, generating light energy – allows scientists to map the magnetic field with a sensitivity never before achieved.

Astronomers have watched in astonishment an auroral 'footprint' up to 1,200 miles (1,900 kilometres) across being generated in the upper atmosphere of Jupiter as a torrent of electrical energy – a current of about 1 million amperes – pours down on the planet from its moon Io. The gases above the cloud tops get heated to more than 5,000 degrees Celsius as a result and the electrical storm fills the sky. Hubble has captured pictures of volcanoes erupting furiously on Io, and scientists think that the volcanic eruptions eject particles that become charged and get trapped by the magnetic field. As Jupiter rotates rapidly under Io the huge magnetic field drags the charged particles round with it, creating an immense sheet of current which, in its turn, modifies the planet's magnetic field. Astronomers using Hubble are working with others controlling the Galileo spacecraft, which went into orbit around Jupiter in December 1995. On the spot, Galileo can measure the type of charged particles present in the aurora, and their energy, and the combination of information will yield better information about this titanic electrical assault by Io on Jupiter, and how Io is managing to generate it.

It is fitting that Jupiter should raise such new, startling questions, even though the planet has been studied for so long, being so obvious and bright in the sky, and therefore so easily observed. Jupiter played a significant role in ending the old Ptolemaic system of astronomy (which put the Earth at the centre of the cosmos) and ushering in today's Sun-centred theory. When Galileo first made his telescope in 1610, he trained it on Jupiter and observed the motion of its four brightest moons as they circled around the planet. Their orbits could not be fitted into the Ptolemaic system at all, and the way was clear for the heliocentric idea to take over. Jupiter is now thought to have at least sixteen moons but the original four observed – Io, Europa, Ganymede, and Callisto –- are still called the Galilean moons.

Jupiter, like the other gas giants, is all too plainly never going to provide a long-term destination for humans, apart from the odd scientist gingerly making observations from a suitable distance. But its moons are a different matter.

There are sixty-one moons identified to date around planets in our solar system, and some of them (including, of course, our own Moon) seem highly likely to be the target of future manned space expeditions aimed at mining, scientific exploration, or even colonisation. So there was great excitement when, in 1995, astronomers announced first that Hubble had detected an oxygen atmosphere around Europa, and then that it had detected ozone on Ganymede.

Previously, oxygen had been found only in the atmospheres of three planets: Earth, Mars and Venus. But, sadly, the discovery of oxygen does not mean that Europa might harbour life. The atmosphere is unbelievably tenuous; although it may extend more than 125 miles (200 kilometres) out into space, the pressure at the surface is vanishingly small compared to that of Earth. Europa is also icy and frigid, with a temperature of roughly minus 145 degrees Celsius. Despite this low temperature the effect of dust and charged particles hitting the surface, together with the meagre sunlight, creates some water vapour which then breaks up chemically to release its constituent hydrogen and oxygen. Hydrogen is so light it quickly escapes into space, but the heavier oxygen stays behind. Europa is approximately the size of Earth's Moon, but looks – and is – very different. The surface is strangely smooth, and made up of solid water ice. Hubble images show mysterious dark cross-hatched marks, giving the satellite what has been termed a 'cracked eggshell' appearance. There is speculation that heating from tidal movement prompted by Jupiter's gravity may mean that there is a hidden ocean of liquid water hidden under the surface crust. Ganymede's ozone is formed by a similar process, but the ozone levels are so tiny that Ganymede is not considered to have an atmosphere at all.

Saturn too has an interesting moon, the giant Titan, which is only a little smaller than Mars. Scientists were able to use the Hubble telescope to make the first maps of the surface of the satellite. Titan is the only body in the solar system, other than Earth, that may have oceans and rainfall on its surface, although the wet stuff is actually ethane-methane rather than water. Scientists suspect that Titan's current state may mirror conditions on Earth billions of years ago, before oxygen began to be pumped into the atmosphere, helping to trap heat energy radiated from the Sun.

Titan's atmosphere is thick (about four times

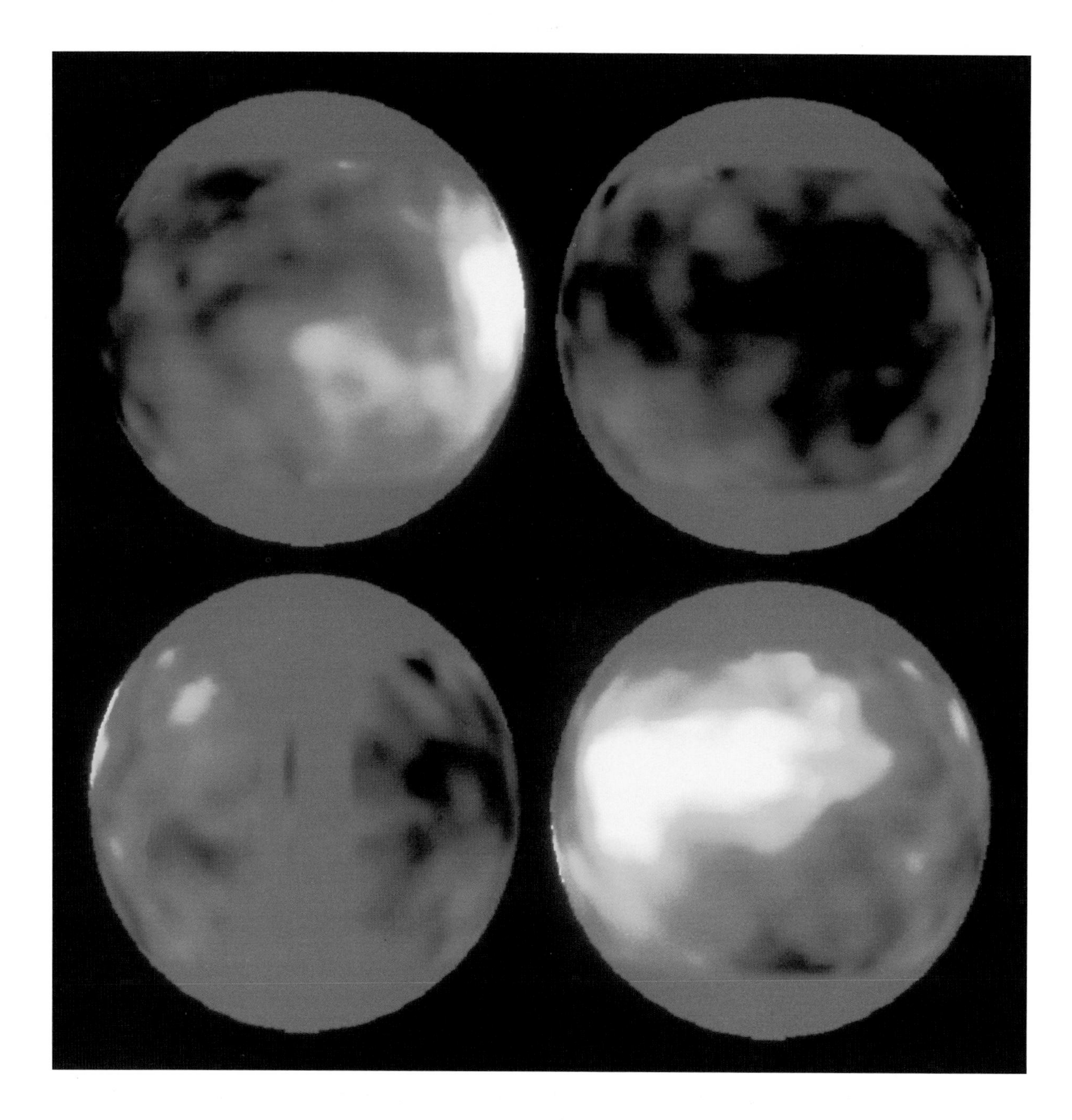

Four faces of Titan, the enormous moon of Saturn, show the first detailed views of its surface. One image (top right) shows a bright area some 2,500 miles (4,000 kilometres) across, which may be solid ground rising out of the oceans of ethane-methane. Titan is bigger than the planet Mercury, and may reflect conditions as they were on Earth billions of years ago.

Credit: Peter H. Smith (University of Arizona), and NASA

denser than Earth's) and is mainly nitrogen (like Earth's) laced with such substances as methane and ethane. When the methane is struck by sunlight, it produces hydrocarbons, generating a soupy, orange haze which we call smog when it blights our cities. The haze was impenetrable to the Pioneer and Voyager spacecraft that flew past Titan in the late 1970s and 1980s, but is brushed aside by Hubble's

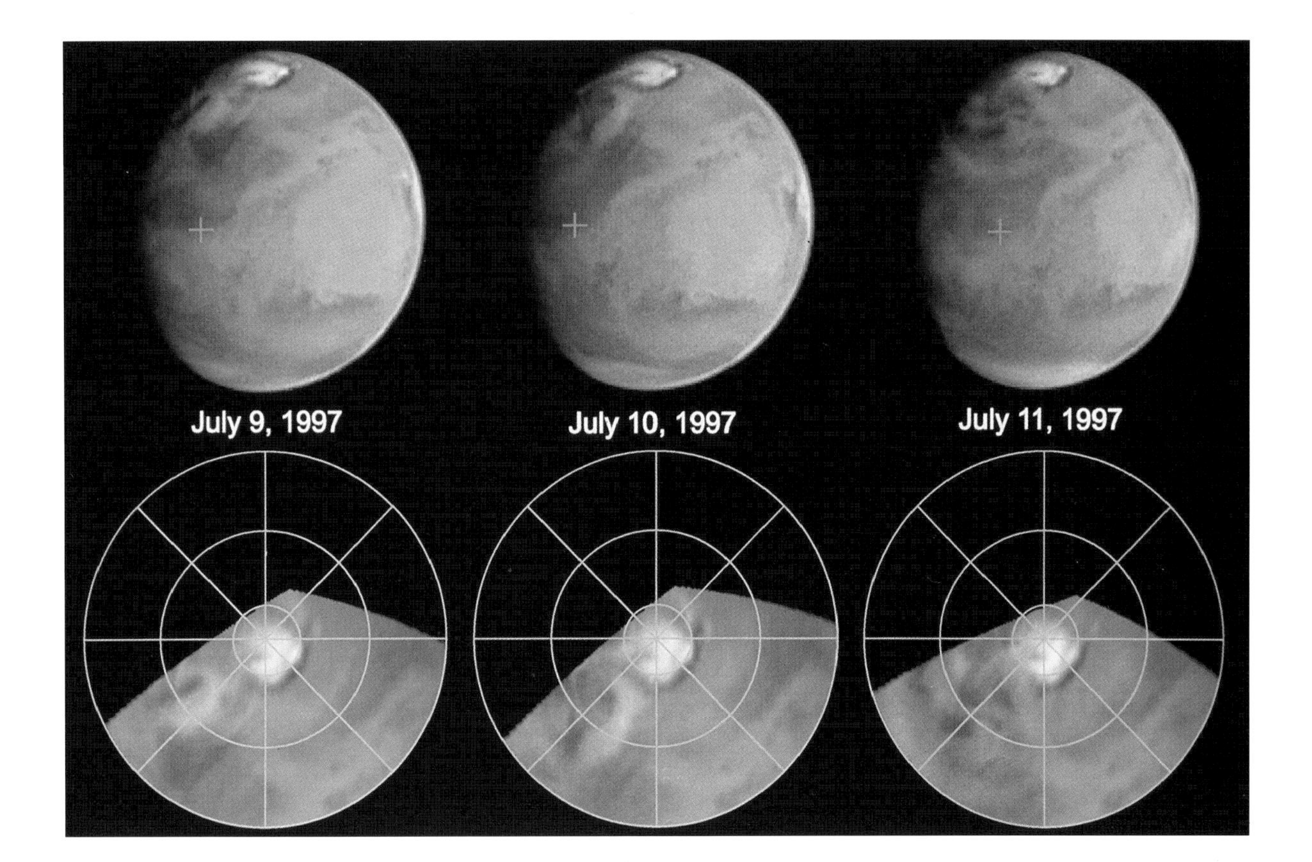

Mars has weather too, as these Hubble images taken over three successive days demonstrate. Bright clouds of water-ice swirl from west to east, while a northern polar dust-storm visible in the first picture rapidly dies away. Detailed reports like these are proving invaluable for the teams directing planetary probes like the Pathfinder mission, whose landing site is shown as a green cross on the Red Planet's surface.

Credit: Steve Lee (University of Colorado), Phil James (University of Toledo), and Todd Clancy (Space Science Institute)

ability to view infra-red light. The telescope found light and dark areas, confirmed that at least some of the surface is solid, and traced out a bright spot about the size of Australia on the surface. Further studies may reveal whether the scientists are looking at mountains, craters, oceans or meteor impact sites.

Again, Hubble will work in partnership with a local unmanned spacecraft, in this case the Cassini mission, which is due to arrive in the Saturn neighbourhood in 2004. Part of Hubble's mission will be to map out the local weather and scope out the best landing spots for a probe that Cassini will parachute on to the moon's surface.

By then, the Hubble teams should be old hands at this kind of mission, having cut their teeth on the Pathfinder mission to Mars in 1997. The usefulness of Hubble's role as weather station was first made clear when it gave the Pathfinder team a better understanding of the weather on Mars than had ever before been achieved. Hubble revealed that the Red Planet is now cooler, drier, and cloudier – but more clear of dust – than was the case two decades ago when the unmanned Viking spacecraft touched down on Mars in the mid-1970s. Ground-based telescopes can get only fleeting glimpses of Mars when it is at its closest approach to Earth. The Viking and Mariner space probes provided plenty of on-the-spot data, but only for an instant of time.

Hubble is able to observe the planet almost continuously, thus tracking climate change and the normal variation of the weather with the changing seasons on Mars.

The information from Hubble was vital to the success of the Mars Pathfinder and the Mars Global Surveyor missions. Pathfinder landed on the surface of Mars in July 1997 – the first spacecraft to do so for more than twenty years – and the Global Surveyor went into orbit around the planet a couple of months later, in September. Hubble supplied data for both missions on the state of the Martian atmosphere, a critical variable because both craft needed to use the resistance of Mars' thin atmosphere for braking and manoeuvring. Variations in temperature, possibly brought on by huge dust storms, can change the density of the planet's atmosphere, thus affecting the point at which air-braking becomes effective. Knowledge of the weather is critical to planning future space missions to Mars, including, possibly, human colonisation.

According to Steven Lee of the University of Colorado in Boulder, one of the researchers who has been using Hubble to study Mars, there has been a global drop in temperature since the time of the Viking landings. Dr Lee stressed how the new data emphasised 'the need for continuous monitoring of Mars. Space probes provided a close-up look, but it's difficult to extrapolate to long-term conditions based upon these brief encounters.' Just after the Viking craft landed, two major storms churned up the Martian atmosphere, leaving fine motes of dust scarcely bigger than smoke particles hanging in the air. The Sun warmed up these dust particles, creating a heat haze that kept the Martian atmosphere hotter than planetary scientists now think is normal. That dust has now settled, and colder conditions are the norm.

Mars is one and a half times as far from the Sun as the Earth is, which means that it receives less than half the heat that shines down on us. Although it rotates on its axis at almost the same rate as the Earth (a Martian day is just forty minutes longer than a terrestrial one) its orbit is so much bigger that a Martian year is nearly twice as long, since the planet takes almost 686 Earth days to go round the Sun. Like the Earth, Mars' axis of rotation is tilted at an angle to its orbit, so that the planet has seasons. But because its orbit is more elliptical than the Earth's the lengths of the seasons are unequal. Spring in Mars' northern hemisphere lasts 199 Earth days, whereas autumn lasts only 147 days. The Martian atmosphere consists almost entirely of carbon dioxide, but this is too thin to retain much of the Sun's heat, so there is little in the way of a greenhouse effect on Mars. In fact, the planet is so cold that its polar caps do not consist just of ordinary water ice, as here on Earth. Carbon dioxide gas from the atmosphere freezes solid as well, so that the ice caps are a mixture of frozen water and 'dry ice'. The frozen carbon dioxide can turn straight from solid to gas without going through a liquid phase, in a process known as sublimation. When it is spring in Mars' northern hemisphere, the dry ice component of the northern ice cap sublimates quickly, so the cap shrinks down to the much smaller permanent cap of frozen water. The differences are dramatic: the dry ice cap can extend to as much as 1,000 miles (1,600 kilometres), whereas the permanent cap of frozen water is thought to be only about 60 miles (96 kilometres) across.

By comparing successive Hubble pictures taken in the Martian spring during 1996, planetary scientists found that the edge of the polar cap had retreated north by about 120 miles (190 kilometres) in only a month. Mars' weather is created by the large differences in temperature between the ice caps (which, being white, reflect the Sun's heat back into space and stay very cold) and the

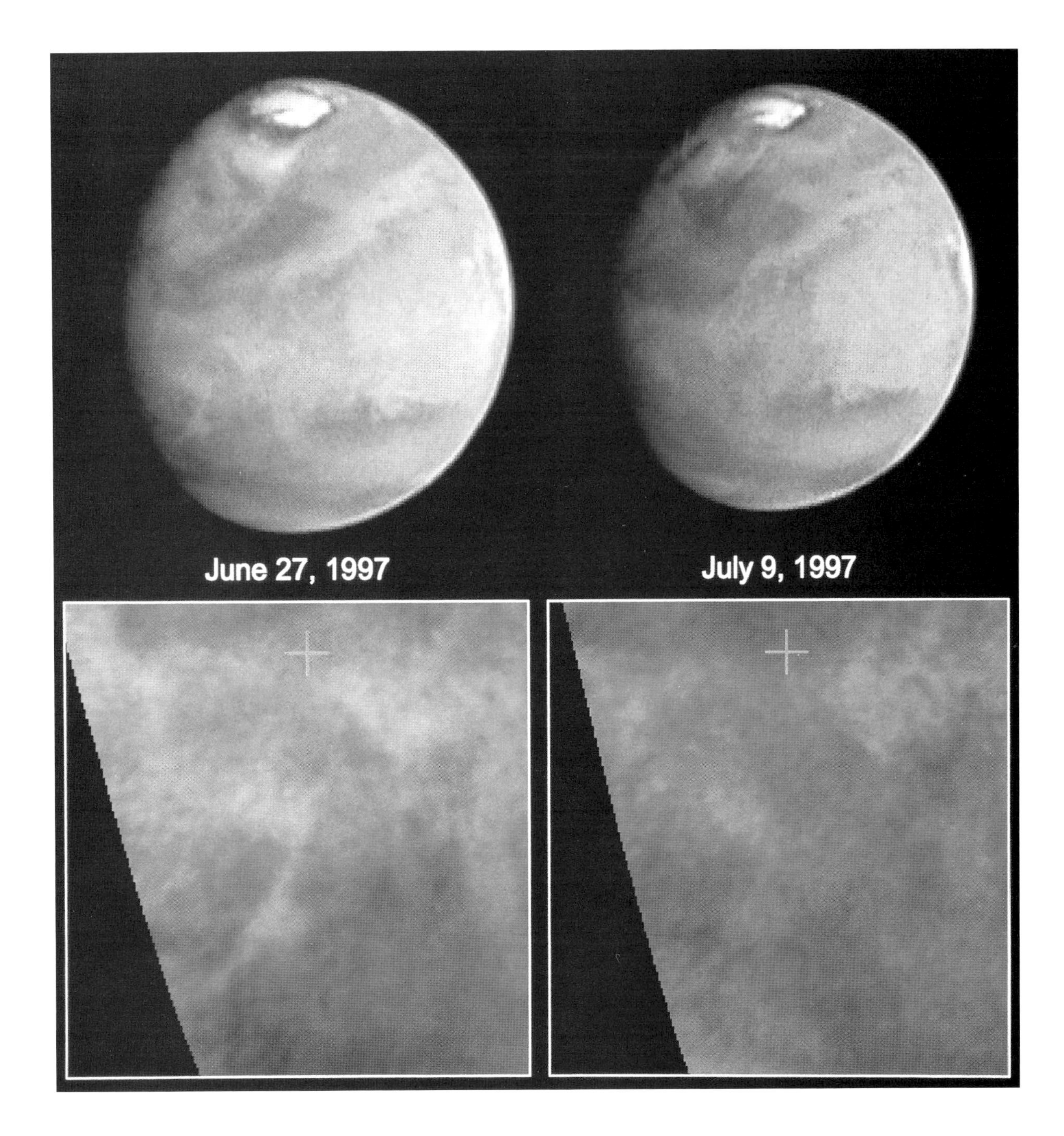

darker-coloured regions immediately to the south, which will warm as they absorb the faint heat of the distant sunlight. The resulting turbulence stimulates the immense dust storms which appear to be a regular feature of the weather on Mars, even though today's storms are smaller than those of the 1970s.

One of the greatest threats to surface exploration of Mars are the vast dust storms that can blow up and die down rapidly. Photographs like these, showing a dust storm blowing itself out to the south of the Pathfinder landing site, were used by the mission team to help make a safe planetfall.

Credit: Steve Lee (University of Colorado), Phil James and Mike Wolff (University of Toledo), and NASA

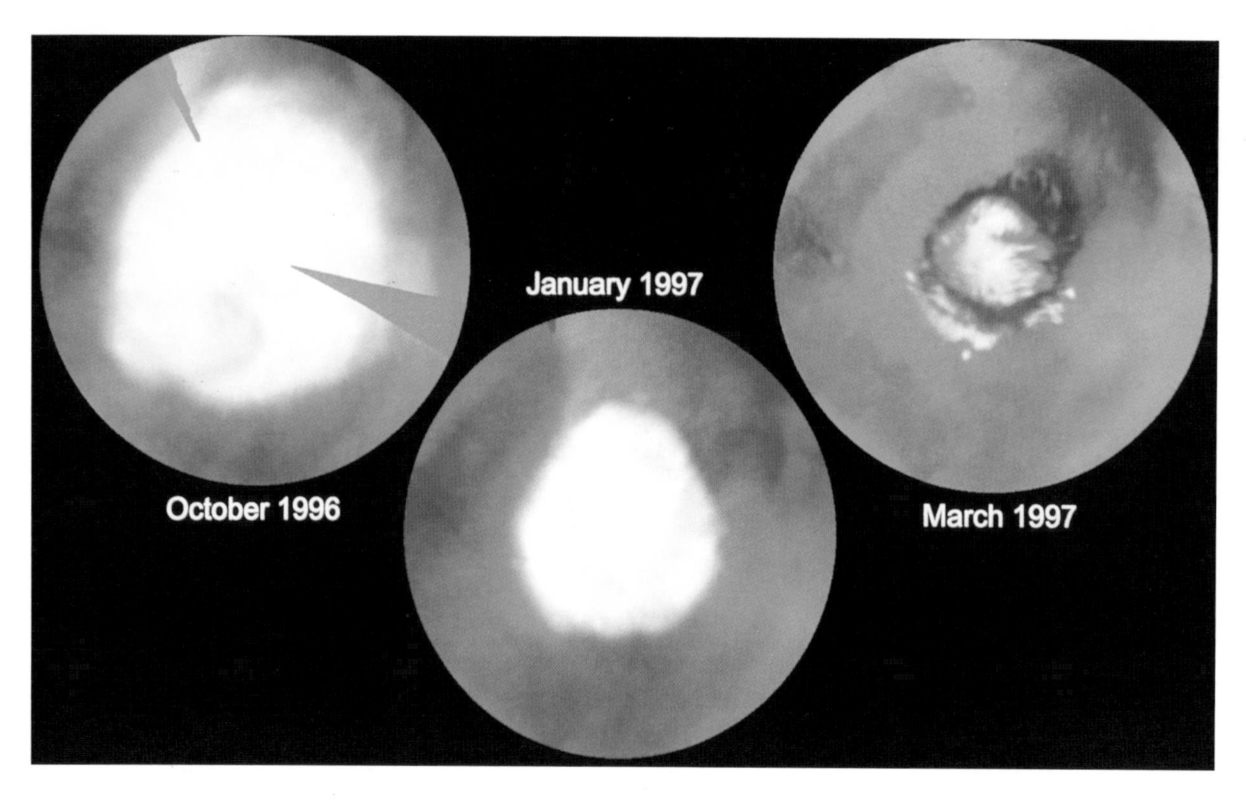

Above: The retreat of Mars' polar ice cap with the march of the seasons.

Credit: Phil James (University of Toledo), Todd Clancy (Space Science Institute, Boulder, Colorado), Steve Lee (University of Colorado), and NASA

Below: Some banding patterns are visible in these images of a planet as distant as Neptune. Vigorous activity can also be seen at the pole.

Credit: L. Sromovsky (University of Wisconsin), and NASA

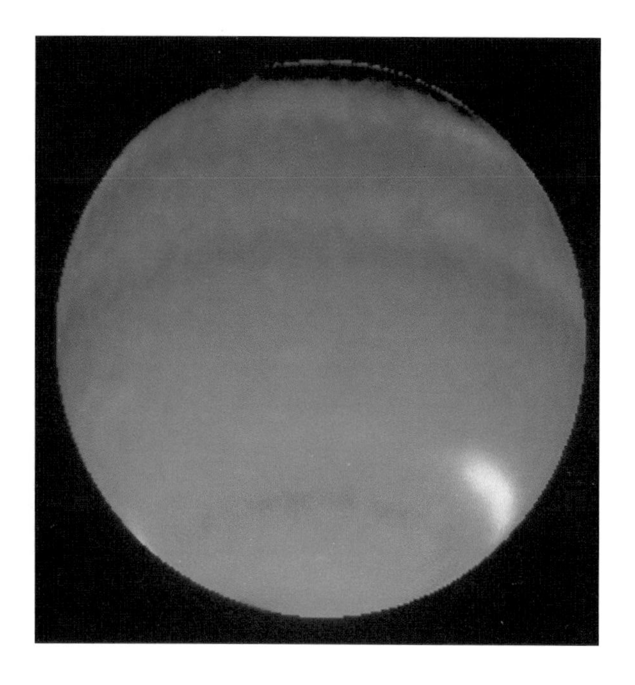

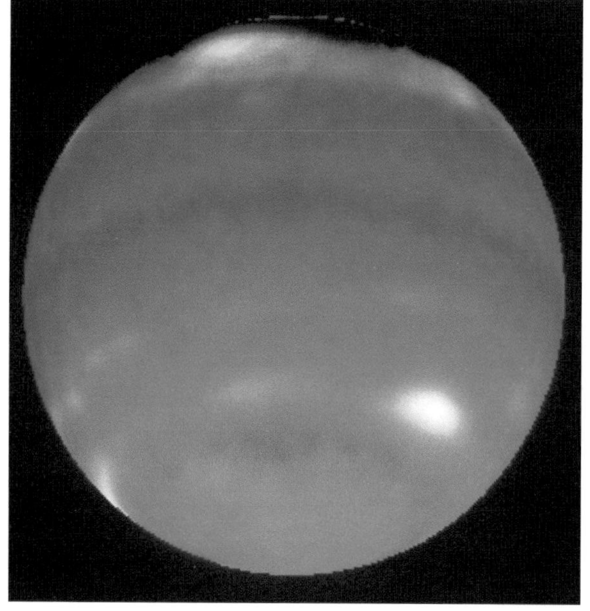

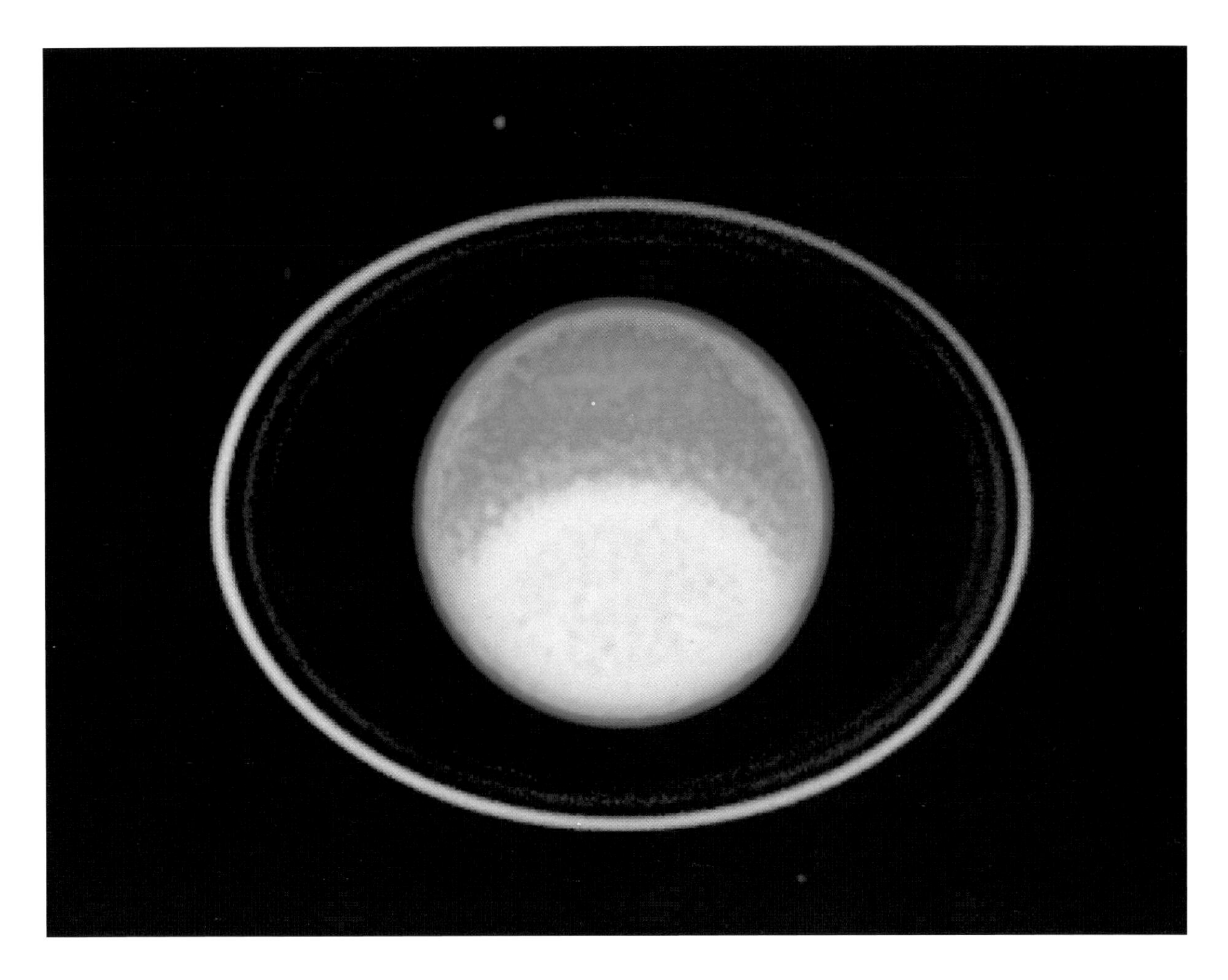

Infra-red filters allowed Hubble to take this picture of the atmosphere of Uranus, which is mostly hydrogen with methane traces. The red at the planet's edge is a thin haze found at high altitudes, which can be seen only when concentrated at the edges of the disk. The planet's rings are also visible.

Credit: Erich Karkoschka (University of Arizona Lunar and Planetary Laboratory), and NASA

As the time for Pathfinder's landing approached, NASA controllers were getting frequent updates on the Martian weather picture, thanks to Hubble's pictures. A fortnight before the landing, a huge dust storm rose up 600 miles (960 kilometres) south of the landing site. Although this blew itself out before the landing, another storm arose on the borders of the north polar cap some 1,200 miles (1,900 kilometres) north of the site. Both were carefully monitored by the Hubble scientists. The changes in the behaviour of the dust and the clouds in the atmosphere have shown that the Martian weather can shift very rapidly.

By contrast, the gas giant Uranus has weather that alters almost imperceptibly over decades. When it was visited by the Voyager space probe a decade ago, the pictures revealed an apparently featureless disk a bland blue in colour with a slight greenish tinge. It would be fair to say that, apart from a die-hard band of dedicated Uranus-watchers, these were pictures that did not exactly set the hearts of the world's scientific community all a-flutter. But when Hubble was turned to look at Uranus, it saw something quite marvellous, something that occurs only once in a human lifetime: springtime on Uranus.

The Uranian year lasts for eighty-four Earth years, and therefore lasts for an entire human generation. Hubble spotted wisps of cloud forming in the upper atmosphere, the first time any such features had been confirmed in recent times. Astronomers were taken by surprise; Uranus is so distant from the Sun – twenty times as far as the Earth – that very little heat must reach it. But Hubble has shown that clouds are gathering, and that the weather is slowly changing. In a lecture to Britain's Royal Astronomical Society in 1996 Dr Heidi Hammel of the Massachusetts Institute of Technology speculated that the reason Voyager took such disappointing pictures of the planet 'just happened to be an accident of timing. Uranus has times when it is active, and that depends on what season it is. We're not used to thinking about seasons that are twenty years long.'

Uranus was first discovered in 1781 by the British astronomer William Herschel. A century later, astronomers in the 1890s reported that they had observed clouded bands on Uranus, but such clouds had not been observed since. Now the images of the Hubble telescope, which is able to picture the atmosphere of the planet developing over time are 'rapidly changing our perception of these planets and how stable they are', Dr Hammel said. Once again it is Hubble's ability to watch the planets over a sustained period of time, rather than snatching a fleeting glimpse in passing as Voyager 2 had to do in 1986, that is making all the difference to planetary science. As springtime on Uranus will last at least a decade or so longer, astronomers will have to have patience and persistence if they are to understand more about this particularly enigmatic planet.

Uranus is the 'tilted' planet of the solar system. The axis about which it rotates virtually lies in the same plane as its orbit, so it appears to corkscrew its way round the Sun, rather than spinning like a top as all the other planets do. Why Uranus should be tilted in this way is not known, but it seems likely that it was knocked over in a collision with a body at least half the size of Earth at some stage during its formation. Uranus' rings and moons orbit the planet's equator, so they must have been formed after the impact, or else they would be rotating in the orbital plane – effectively going round the north and south poles.

Uranus has five largish moons and at least ten others, which whizz rapidly through the skies above the planet's surface in complicated patterns. The smallest and closest of the planet's major satellites, Miranda, bears witness to some cataclysm early in its history. According to pictures sent back by the Voyager mission, it appears to have been torn apart by a huge impact and then put back together again. Even though Miranda is only 310 miles (500 kilometres) across it is riven by cliffs, craters and canyons, some of which are ten times deeper than the Grand Canyon on Earth.

Although Uranus is classed as a gas giant planet like Saturn and Jupiter, with a diameter about four times that of Earth, the gas does envelop a central rocky core roughly the same size as Earth. The collision that tilted Uranus would have knocked lots of bits out from the embryonic planet, and these would have been very different in character from the original material of the proplyd that came together to form the planet in the first place. So astronomers had a shrewd suspicion that the composition of the moons and the eleven rings of dark dust round Uranus might differ from those round the other gas giants. And when they looked at the light from the Uranian rings and the Uranian satellites through a spectroscope, they found it did indeed differ from the light reflected from the moons and rings around Saturn and Jupiter.

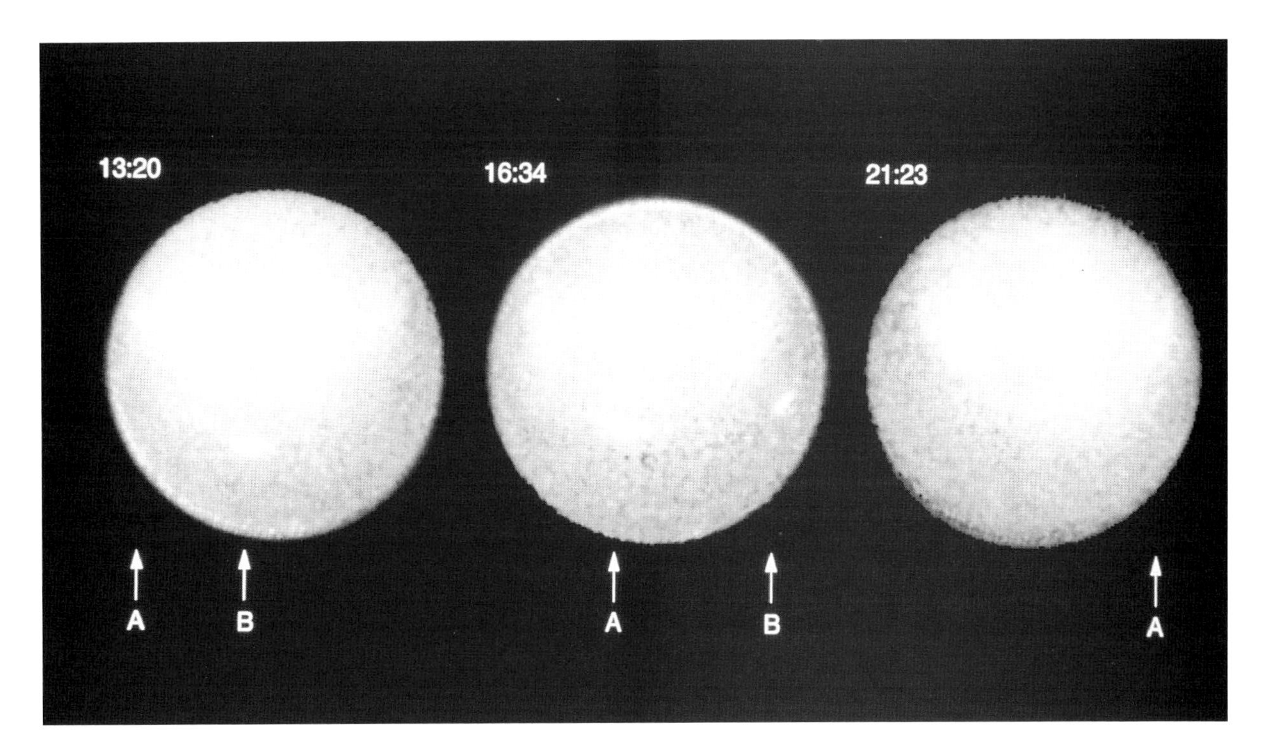

Above: These three shots of Uranus were taken over 8 hours, and the movement of two bright cloud formations in the southern hemisphere gives away the speed at which Uranus rotates on its axis – the day is just 7 hours and 14 minutes long.

Credit: Kenneth Seidelmann (U.S. Naval Observatory), and NASA

Below: This view of Uranus' rings and five of her moons was taken when the planet was 2.8 billion kilometres (1.7 billion miles) away. The three images of each moon are due to the picture being composed of three separate images, taken about 6 minutes apart – showing how fast the moons travel.

Credit: Kenneth Seidelmann (U.S. Naval Observatory), and NASA

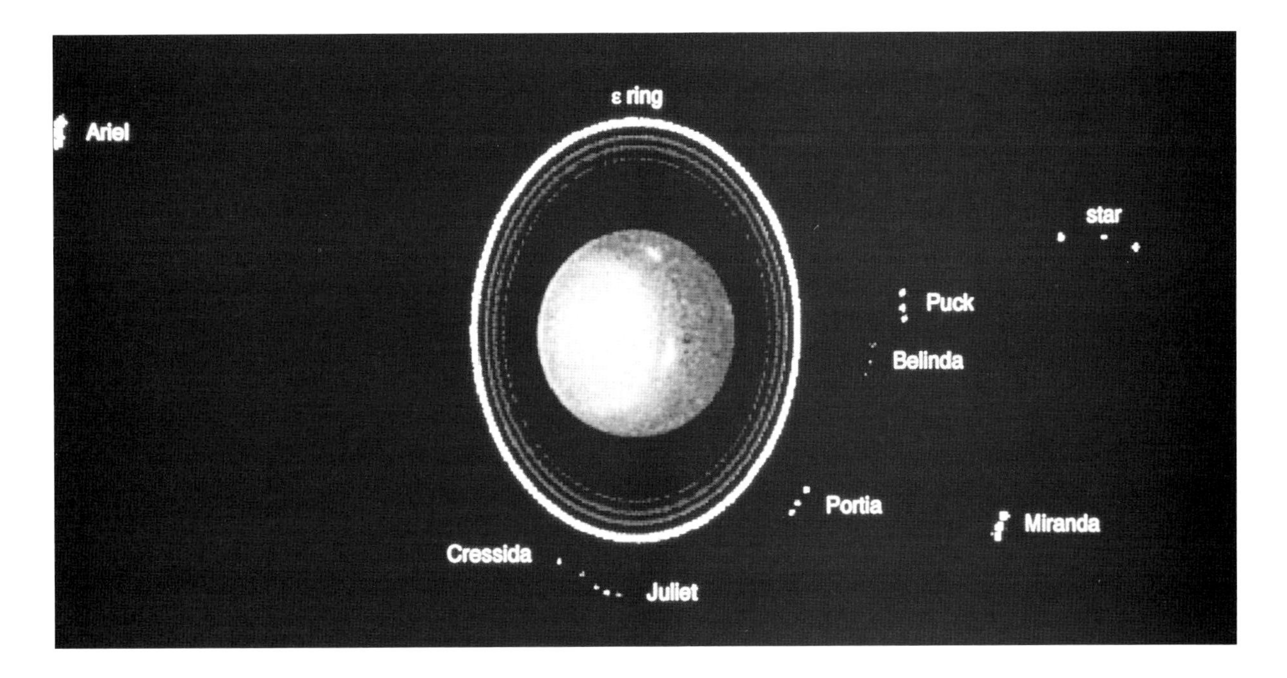

The dark material that we see in the Jovian and the Saturnian systems is actually slightly red, and astronomers believe that the material in those systems consists of complicated compounds of carbon. However the dark material in the Uranian system has a different spectral composition: it has about the same brightness across the whole visible portion of the spectrum, so it appears to be rather grey. Usually, grey is used as a synonym for dull and uninteresting, but for once it is the very greyness of the Uranian system that makes it fascinating and different. The composition of the satellites and rings that encircle Uranus is now one of the most pressing questions facing planetary scientists. Voyager was not equipped to investigate that issue. Hubble is in a position to come up with some answers.

As it is with Pluto, the smallest and strangest of all the planets. Pluto is unique in that it is neither a rocky planet like Earth or Mars, nor a gas giant like Jupiter or Uranus, but is instead an enormous iceberg covered in a coating of frozen methane. Pluto was discovered only in 1930 and its biggest moon, Charon, was found as recently as 1978. Charon is more than half Pluto's diameter, and is a mere 12,000 miles (19,300 kilometres) or so from the planet; by comparison, if our own Moon were that close, it would appear as large in the sky as an apple held at arm's length. They really form a double planet system, and some astronomers speculate that Charon may have been born as a result of a head-on collision between Pluto and some other object,

Pluto and her moon Charon are effectively a double planet. Seen by ground-based telescopes, the two are indistinguishable; indeed, Charon remained undetected for almost 50 years after the discovery of Pluto. When photographed, the pair were 30 times as far away from Earth as we are from the Sun.

Credit: Dr. R. Albrecht (ESA/ESO Space Telescope European Co-ordinating Facility), and NASA

Hubble was able to offer scientists the first ever shots of the surface of Pluto, revealing to their trained eyes polar caps, dark and light spots, some mysterious linear markings – altogether a dozen discrete provinces, none of which had ever been seen before.

Credit: A. Stern (South-West Research Institute), M. Buie (Lowell Observatory), NASA and ESA

rather as the Earth-Moon system is thought to have formed. Debris from the collision, so the theory goes, orbited Pluto and gradually coalesced into Charon.

Pluto's orbit brings it in (comparatively) close to the Sun every two and a half centuries, as is the case just now. When it comes closer to the Sun, Pluto behaves almost like a comet by warming up and losing some of its atmosphere into space. But it is much bigger than any known comet and must surely be one of the last survivors of the 'ice dwarfs' – larger, icy objects that were part of the proplyd out of which the rocky planets and the gas giants condensed. Pluto is perhaps not so much the last of the planets as the first of the objects of the Kuiper Belt, a sort of super-comet. Astronomers think that the majority of these ice dwarfs were sent spinning out of the solar system hundreds of millions of years ago by massive collisions with other objects, although some may have been 'saved' into stable orbits by being captured by the gravitational pull of larger objects. Thus Pluto was captured by Neptune's gravity, and there are suggestions that Triton, one of Neptune's moons, might be another ice dwarf captured by its giant neighbour and pulled into closer orbit.

However, Hubble has found features on Pluto which suggest that it is not really a close relative of Triton. Although Pluto is small – just two-thirds the size of our own Moon – and very distant, the

European Space Agency's Faint Object Camera on board Hubble has taken the first pictures showing any details on the surface of Pluto. (This feat is equivalent to spotting smudges on a baseball 40 miles/65 kilometres away!) To the untrained eye, these look like vague smears of grey shading into white, but the very fact that the Hubble images reveal much more surface variety on Pluto than on Triton suggests that Pluto has intriguing differences.

The educated eye of the planetary scientist can make out ragged icy polar caps and clusters of bright and dark features that might be impact craters or valleys and basins. Some of the changes in surface contrast are likely to be due to seasonal changes in the distribution of frosts that traverse the surface. As Pluto passes out into the frozen outer reaches of the solar system, it dumps much of its atmosphere back onto the surface as fresh ice. These new frosts are very strange to us – composed of nitrogen, carbon monoxide and methane – but yet again, Hubble is reporting weather, albeit weather a couple of billion miles away.

The planets and moons in our back yard make up an intriguing and splendidly varied bunch, and scientists hope that the years of closer study ahead will equip us with at least some of the answers to the questions that we posed at the beginning of this chapter. We have, in effect, a series of local test-beds close to hand that may end up giving us an understanding of where to look for alien planets, what those planets might be like, and which will be the best candidates for having developed life beyond Earth.

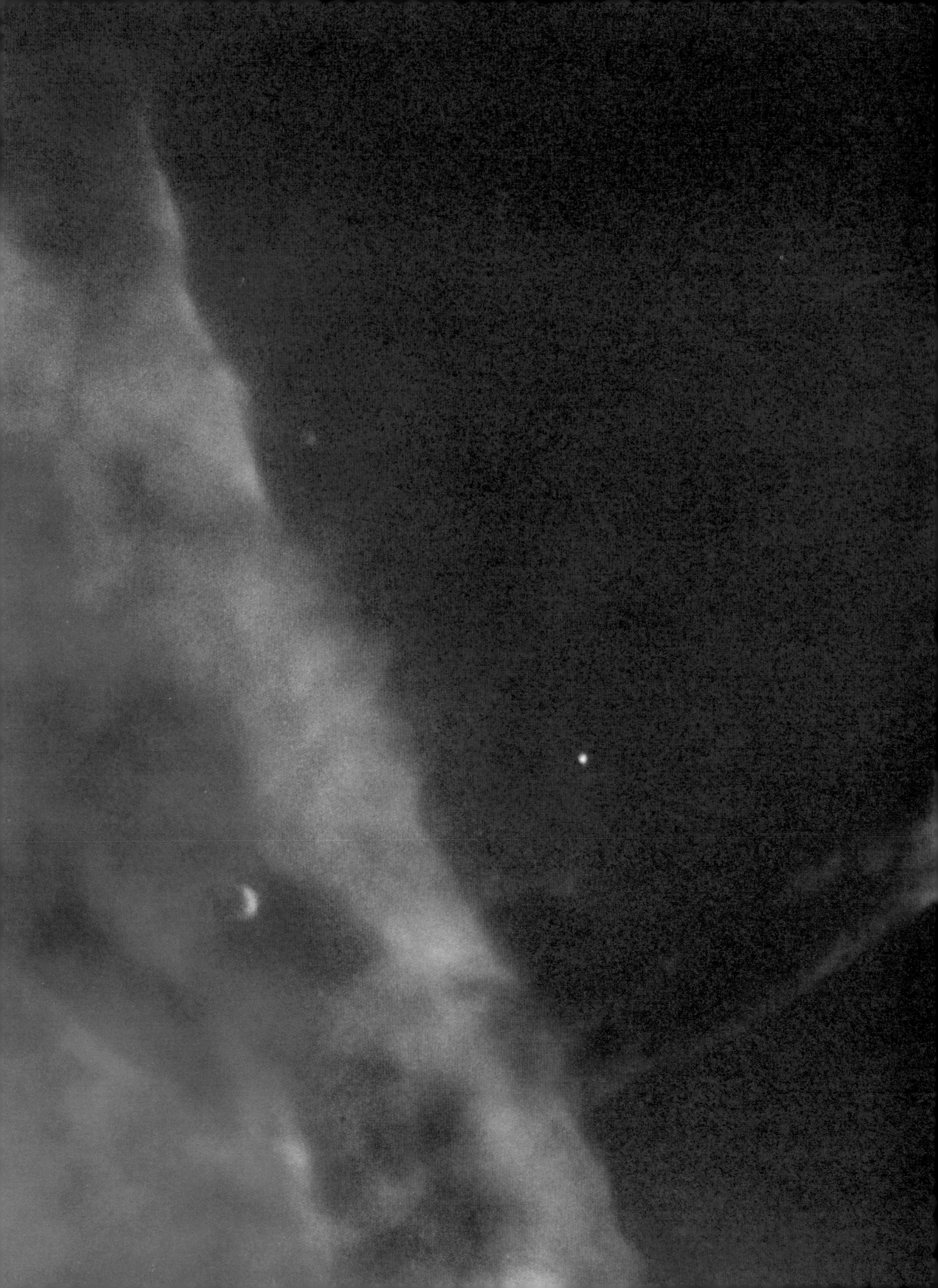

CHAPTER THREE

THE CRADLE OF THE STARS

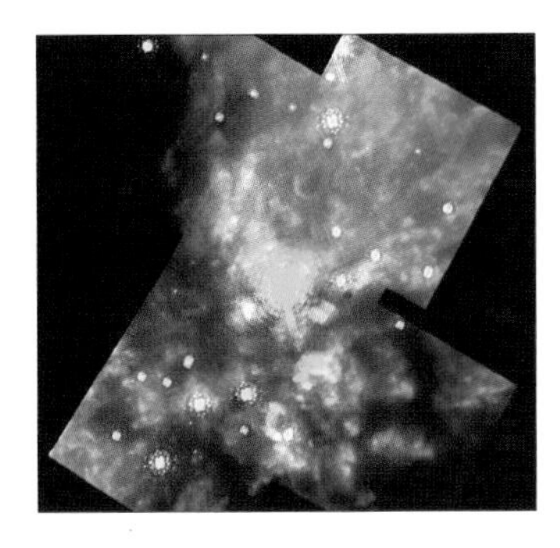

Imagine you are an astronaut of the future taking a fast spaceship on a journey to the edge of our solar system. (It is a journey that the unmanned Pioneer and Voyager probes have already taken and which lasted more than a decade.) First you would pass through the familiar clutter of the planetary orbits, packed with colour and incident provided by gas giants, rocky planets, asteroids, moons, and all manner of flying junk. Then, as Pluto and its constant companion Charon dwindled behind your vessel, you would enter the realm of the Kuiper Belt, with its scattering of dirty icy chunks, the rubble left over from

Opposite: The Orion nebula, birthplace of the stars. This picture shows a close-up of a hypersonic shock wave ploughing through the nebula, which is an enormous cloud of gas and dust. Scientists operating the Hubble Space Telescope trained its instruments on the Orion nebula because they knew that hidden deep inside it they would find young stars at the moment of their creation.

Credit: C. Robert O'Dell, Shui Kwan Wong (Rice University), and NASA

Above: Hubble's ability to see into wavelengths other than just that of visible light – ultraviolet and infra-red – means that it can penetrate obscuring layers of dust to see into the heart of Orion. Here, Hubble's superior vision has found a region of intense starbirth, where the interstellar gas and dust heated by the intense energy is glowing yellow-orange. The bright object is a massive young star (see pp.80-81).

Credit: Rodger Thompson, Marcia Rieke, Glenn Schneider, Susan Stolovy (University of Arizona); Edwin Erickson (SETI Institute/Ames Research Center); David Axon (STScI); and NASA

the building of the system itself. Astronomers have calculated that there cannot be enough material beyond Pluto to form another planet. So what then after the Kuiper Belt? According to the Dutch astronomer Jan Oort, you might find a fresh cloud populated with 100 billion icy, comet-like objects. Despite its immense population, the Oort cloud would be spread across so vast an area that you might not notice any of its denizens as you passed through. And beyond that?

Space.

By now, the life-giving warmth of the Sun would be just a memory. You would have passed the boundary of the 'heliosphere' – the area of influence of our Sun, where the interplanetary magnetic field and the 'wind' of subatomic particles from the Sun dominate – and you would be out into the interstellar medium. Our star would be visible only as one among the many millions shining silently in the utter cold of the void.

Suppose our astronaut decided to visit the nearest available star. Our next-door neighbour is called Proxima Centauri. This is not a bright star; indeed, it is a dwarf star so dim that from Earth it can be seen only with the aid of a telescope. If any telescopes are trained on our neighbour tonight, they will be picking up (rather feeble) light that has taken nearly four years and four months to get here – even though the light has been travelling towards us at 186,000 miles (300,000 kilometres) per second. Were our astronaut to try to reach our closest stellar neighbour, he or she would die of old age before even a fraction of the trip was complete. Using present space flight technology the journey to Proxima Centauri would last more than 100,000 years. Which puts the old human dream of 'journeying to the stars' into a somewhat sobering context.

If you were aboard the ship, you could easily imagine that you were drifting in a nothingness, a complete emptiness between our solar system and the other stars. And you would be wrong; space is not empty, although it may seem that way. Were you able to see them, you would be able to spot an atom (mostly hydrogen) every cubic centimetre or so and a grain of dust floating in every 100,000 cubic metres of space. But the only real landmarks of any size you could see would be the stars themselves.

The stars, then, are apparently the largest and most significant objects in our galaxy. Yet if that were truly so, we would face an unanswerable question: where did the stars themselves come from? For the stars we see in the night sky are younger than the universe itself. Our own Sun, for example, is roughly 4.5 billion years old, whereas the universe is at least twice and probably three times that age. What was there before our Sun?

The answer to these questions springs from the earlier observation that space is by no means completely empty. At least a slight trace of matter – gas, dust – is present everywhere in interstellar space. Because space is so big, even matter that seems incredibly thin and tenuous can end up forming something as dense and powerful as a blazing star. And when the stars die, as we shall see in the next chapter, they return to dust and gas; but with a difference, for there is evolution in our universe, and the ashes of dead stars are no longer made up of the same dust and gas from which they were originally created.

The life-cycle of the star revolves around this exchange of matter into and out of this interstellar medium. It is a process that proceeds on timescales unimaginably long by human standards: millions to billions of years. To understand the birth and death of stars, astronomers need to observe this exchange taking place; and the Hubble Space Telescope has given them a ringside view.

Through the byways of the galaxies drift immense clouds of interstellar dust and gas.

A close-up of a region of the Eagle nebula where young stars are being incubated. The finger-like protrusions at the top of the nebula – each fingertip is larger than our entire solar system – contain clumps of especially dense gas, within many of which are baby stars. The density of the clumps is protecting them from the intense ultraviolet energy produced by nearby stars, which is boiling away the rest of the pillar – a process known as 'photoevaporation'.

Credit: Jeff Hester and Paul Scowen (Arizona State University), and NASA

Although the density of this gas is extremely low, the total quantity of material floating about is huge. About 10 per cent of the material in our own galaxy, the Milky Way, consists of dust and gas between the stars – equivalent to the mass of 10,000,000,000 Suns. In fact, some of this material collects together to form the largest 'objects' in the galaxy, called giant molecular clouds. These clouds are huge; so much so that in many cases it would take a beam of light two centuries to pass from one side of the cloud to the other. These clouds are the cradles of the stars.

The clouds are mainly made up of hydrogen and helium, but in all more than sixty different types of molecules have been detected inside them, including carbon monoxide, water, ammonia, plus other, more complex chemical species such as formaldehyde and alcohol. Like all fogs, they are patchy, fainter in some regions and denser in others. Fogs here on Earth can be concentrated and dissipated by wind; but for these clouds in interstellar space the force at play is not wind but gravity, the universal attractant, which acts to make the dense patches still denser.

By and large, molecular clouds are found in the younger areas of a galaxy where starbirth is happening today. There are many clouds in the spiral arms of our local galaxy, the Milky Way, not far (at least by astronomers' perspective) from ourselves. But before they give birth to the young stars these clouds are cool, and are therefore difficult to spot; so the act of conception – the way in which the clouds condense into stars – is hard to observe. Visible light is, for once, not very useful in trying to observe the birthplaces of the stars, and astronomers need to use other parts of the electromagnetic spectrum. Thus some cloud regions too cool to emit visible light are just warm enough to radiate infra-red heat radiation; in other regions, dust absorbs all the visible light but lets infra-red light through. Where the young stars have already

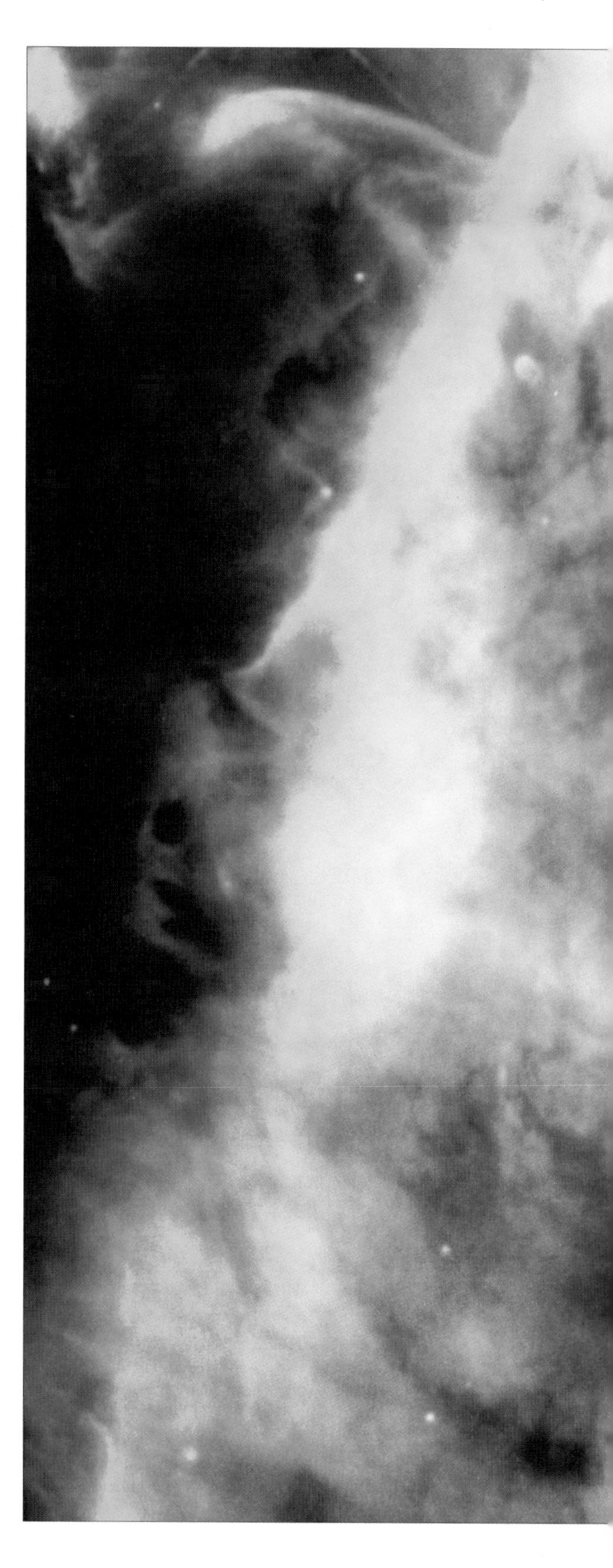

Stars were born in this part of the Orion nebula very recently (300,000 years ago). Plumes of gas surge across the image, material being ejected from newly formed stars. The lumps and blobs visible in the image are young stars surrounded by disks of dust and gas. Most of these disks are about twice the diameter of our own solar system.

Credit: C.R. O'Dell (Rice University), and NASA

begun to shine, in contrast, much of the radiation emitted is in the form of highly energetic ultraviolet light – on the other side of the visible part of the spectrum from infra-red.

Because of the absorbing effects of the Earth's atmosphere, neither infra-red nor ultraviolet light can be observed readily by ground-based telescopes. The Hubble Space Telescope, however, can 'see' into both infra-red and ultraviolet and it is unencumbered by the atmosphere, so the birthplaces of the stars were an obvious target for its observations.

The stellar maternity wards have wonderful, exotic names. The nebula in the constellation of Orion the Hunter, discussed in the previous chapter, is one such cloud where starbirth is taking place today. It is the nearest stellar nursery. Another cradle of stars lies in a lesser-known constellation next to Orion called Monoceros, or the Unicorn. Others have been observed in the Lagoon nebula in the constellation of Sagittarius, and in the Eagle nebula in the constellation of Serpens.

The very early stages of starbirth are slow and ponderous, as gravity encourages the denser patches in the clouds gradually to accumulate more material. Sometimes the process is nudged along by shock waves from a nearby old star as it explodes into a cataclysmic supernova. As more and more material condenses, it begins to collapse under the force of its own gravity. Smaller, more dense clouds start to form, called Bok globules after the Dutch-born American astronomer Bart Bok who first studied them. These dense and dusty clouds can have a mass as much as 2,000 times that of the Sun within a region that measures only three light years across – smaller than the distance between the Sun and Proxima Centauri.

As the globule collapses, its temperature rises a millionfold. The temperature of a giant molecular cloud is around 10 degrees Celsius above absolute zero, but the temperature at which a protostar will ignite is when its core reaches 10 million degrees above zero. At this point the hydrogen atoms begin to fuse together to form helium, releasing vast amounts of thermonuclear energy. Part of the Bok globule explodes like a hydrogen bomb – and a star is born. As the star ignites, the force of the thermonuclear explosion sends shock waves through the rest of the cloud. The ripples can trigger the condensation of other regions into Bok globules and protostars. Like a chain of dominoes, star formation can thus start on one side of a giant molecular cloud and sweep through to the other. ('Sweep' is, of course, used here in the astronomers' parlance; the process may take 10 million or more years.)

While the early process of condensation and collapse may have been ponderous, once a star catches ablaze with its internal nuclear fires, starbirth becomes uproarious and chaotic. The upgraded Hubble Space Telescope now carries a 'near infra-red camera and multi-object spectrometer' (NICMOS) which can 'see' heat radiation from objects which, although warm, are not yet hot enough to glow with visible light. NICMOS can thus partially see through dust clouds which absorb visible light. Its images and the scientific data drawn from them have revealed the true anarchy of starbirth, with disordered arrays of clumps, bubbles, and knots of material shooting out, along with streams of hydrogen travelling at 1 million miles (1.6 million kilometres) per hour. These 'molecular bullets' are colliding with other, slower-moving material causing a bow wave, like a ship ploughing through water.

The intense starlight of the vigorous young stars heats the gas and dust to an extent that, in one NICMOS image of the Orion nebula, the interstellar dust appears to glow yellow-orange. The brightest object in the image is a massive young star, first discovered in 1969 by Eric Becklin and Gerry Neugebauer of the California Institute of Technology, when they looked at the heat radiation (infra-red light) coming from the Orion nebula. This star started to shine out at about the time early modern humans first appeared on the plains of Africa, a few tens of thousands of years ago; so it is a very young star indeed. But the equipment on the Hubble Space Telescope, whose infra-red vision allows it to peer through the murk, has found blue fingers of radiation being emitted by molecular hydrogen, indicating that there are even younger stars, still obscured by the dust, shooting out violent bursts of material. The trained eye of the astronomer can also pick out crescent-shaped bow shocks generated by the flowing material.

Eventually the new young stars become visible as the intense ultraviolet light they produce boils away the rest of the dust and gas surrounding them, a process called photoevaporation. Stellar growth is thus self-limiting, for the vast outpouring of energy from nuclear fusion within a young star disperses the remains of the cloud of material which might otherwise have been absorbed to allow the star to grow even bigger.

Hubble has caught sight of just such an energy 'hurricane' ripping through the Eagle nebula in the constellation Serpens, 42 million billion miles (68 million billion kilometres) away. (The events it is showing us took place 7,000 years ago; as they looked at the Hubble Space Telescope's images, taken in late 1995, astronomers were analysing pictures of starbirths that took place at least 1,000 years before human civilisation started the systematic study of the stars.)

The Eagle nebula as seen in the Hubble pictures is a massive 'stalagmite' of cool dust and gas towering up from a huge cloud of hydrogen, like some cosmic stalagmite growing from the floor of a cave. The scale of the pictures is staggering. Each of the columns or gaseous towers is several light years long – perhaps the distance between Earth and Proxima Centauri. If you were to build a spaceship capable of travelling at the speed of light (which according to Einstein is impossibly fast) it would take you more than a year to get from the bottom of the gas pillar in the picture to the top. The images are achingly beautiful, but astronomers were excited by more than just their aesthetic appeal. The gas inside the columns is dense enough to collapse under its own weight, forming young stars that continue to grow as they accumulate more material from their surroundings. For the first time, astronomers could see clearly some of the earliest stages in the process of star formation, including the ghostly streamers of gas being boiled off the columns by photoevaporation.

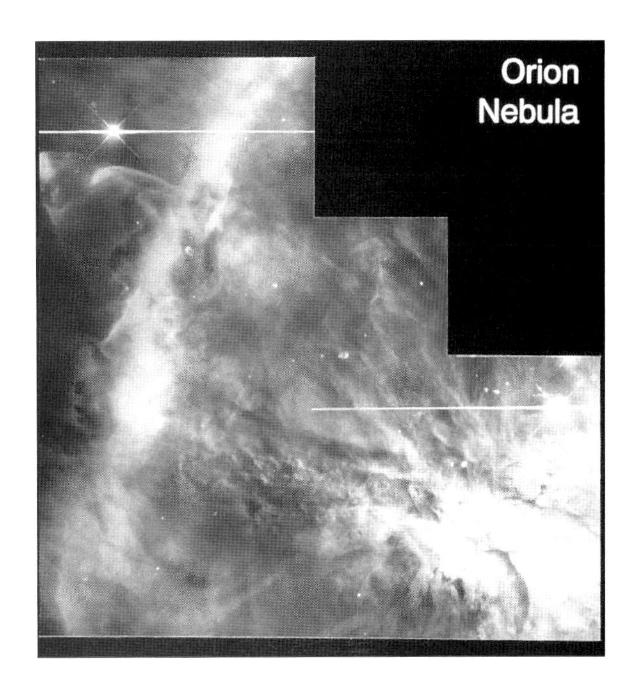

The Great Nebula in Orion. Images from Orion that Hubble takes today have taken 1,500 years to travel here at the speed of light. This is how, in effect, the Hubble Space Telescope is acting as a 'time machine', showing us events that took place long ago.

Credit: NASA

Opposite: Ghostly streamers of gas flowing out of this cloud of cool gas show how the ultraviolet energy given out by nearby hot stars (out of the picture) is gradually stripping away the structure of the pillar, which is a close-up of the starbirth region seen above.

Credit: Jeff Hester and Paul Scowen (Arizona State University), and NASA

Above: Arguably the most famous photograph ever taken by Hubble. This image of vast pillars of gas in the Eagle nebula, 7,000 light years away from Earth, shows a region where stars are being born as gas and dust compacts under its own gravity, becoming dense enough to trigger the nuclear engine that powers a star. The pillars themselves have been left behind as photoevaporation has 'boiled' away surrounding, less dense gas – rather like rocky pillars left standing in deserts after erosion has worn away the softer rocks that once surrounded them.

Credit: Jeff Hester and Paul Scowen (Arizona State University), and NASA

Not all the gas evaporates at the same rate, and as the outer layers disperse they reveal the denser globules – so-called 'evaporating gaseous globules' or EGGs – within which the new stars are forming. Jeff Hester of Arizona State University, whose team took the images, said: 'It's a bit like a wind storm in the desert. As the wind blows away the lighter sand, heavier rocks are uncovered. But instead of rocks, the ultraviolet light is uncovering the egg-like globules of gas that surround stars that were forming inside the gigantic gas columns.' Some of these EGGs appear just as tiny bumps on the surface of the column or, if they have been uncovered more fully, as fingers of gas protruding from the larger cloud.

'This is the first time we have actually seen the process of forming stars being uncovered by photoevaporation,' Dr Hester said. 'In some ways, it seems more like archaeology than astronomy. The ultraviolet light from the nearby stars does the digging for us, and we study what is unearthed.'

A similar picture is revealed looking in towards the centre of the Milky Way. In the constellation of Sagittarius the Archer, which lies in the direction of the centre of our galaxy, sits the Lagoon nebula, 5,000 light-years away. The nebula was first sighted a long time ago, in 1747. It is cut by a dark cloud and it is from this feature that it gets its name.

Hubble here also found the now familiar glowing clouds of dust and gas, shining with an eerie inner light, heated by radiation from stars as yet invisible. One central hot star, Herschel 36, is visible. Present too are the structures of starbirth: the Bok globules, the bow shocks around new stars, the wisps and jets in the clouds of dust and gas. Hubble saw also a pair of what looked like tornadoes, funnel-like structures resembling twisted ropes. Each of the interstellar 'twisters' are about half a light-year long, i.e. about 3 million million miles (5 million million kilometres) long. The blue mist in the image is photoevaporation of the clouds. And here too the effect of the intense radiation from young stars is acting like a wind on Earth parting the mists and fog. Indeed, astronomers speculate that the temperature difference between the surface of the clouds – heated by the starlight – and their cooler interiors may generate tornadoes just like the violent disturbances of the atmosphere here on Earth, but on a truly enormous scale.

Hubble has also managed to capture images from the constellation of Monoceros showing some of the later stages of starbirth. Monoceros is a faint constellation adjoining Orion, which contains several nebulae and clusters of stars. Using NICMOS to peer into the infra-red, Hubble focused on the Cone nebula, 2,500 light-years distant from Earth, and identified starbirth taking place there within a star cluster known as NGC 2264. In particular, Hubble picked out a 'family group' of young stars centred on a bright, massive star called NGC 2264 IRS, finding no fewer than six 'baby' stars surrounding it. The massive star appears to have triggered their creation by the release of high-speed particles of dust and gas during the early stages of its formation. Ground-based telescopes are unable to see the mother and its baby stars because the dust and gas obscure them from our view.

Many mysteries surround the birth of the stars because of this omnipresent obscuring dust. In the early 1950s, before the advent of space-based observatories, two astronomers – George Herbig, an American, and the Mexican Guillermo Haro – independently discovered small, faintly luminous objects which, the Hubble Space Telescope has now revealed, represent the final stages in the birth of stars. Named 'HH objects' in honour of their discoverers, more than 300 have now been detected. They are irregular in shape and contain bright 'knots'. Because they were found in regions known to be associated with star creation, they were ini-

tially thought to be recently formed stars but they changed so quickly that astronomers realised the HH objects could not be stars themselves. Over the four decades since the discovery of HH objects, astronomers have pieced together a plausible picture suggesting that each one is a very young star hidden in a cloud of dust. Streaming out from the star is a violent solar 'wind' of subatomic particles

The Lagoon nebula: the 'twisters' described on page 84 are at the top left of the picture. A Bok globule can be clearly seen in the bottom left-hand corner. This area of the nebula is dominated by Herschel 36, the hot star shining near the top of the photograph, whose radiation (along with that of other nearby stars) is responsible for most of the movement and activity visible in the image.

Credit: A. Caulet (ST-ECF, ESA), and NASA

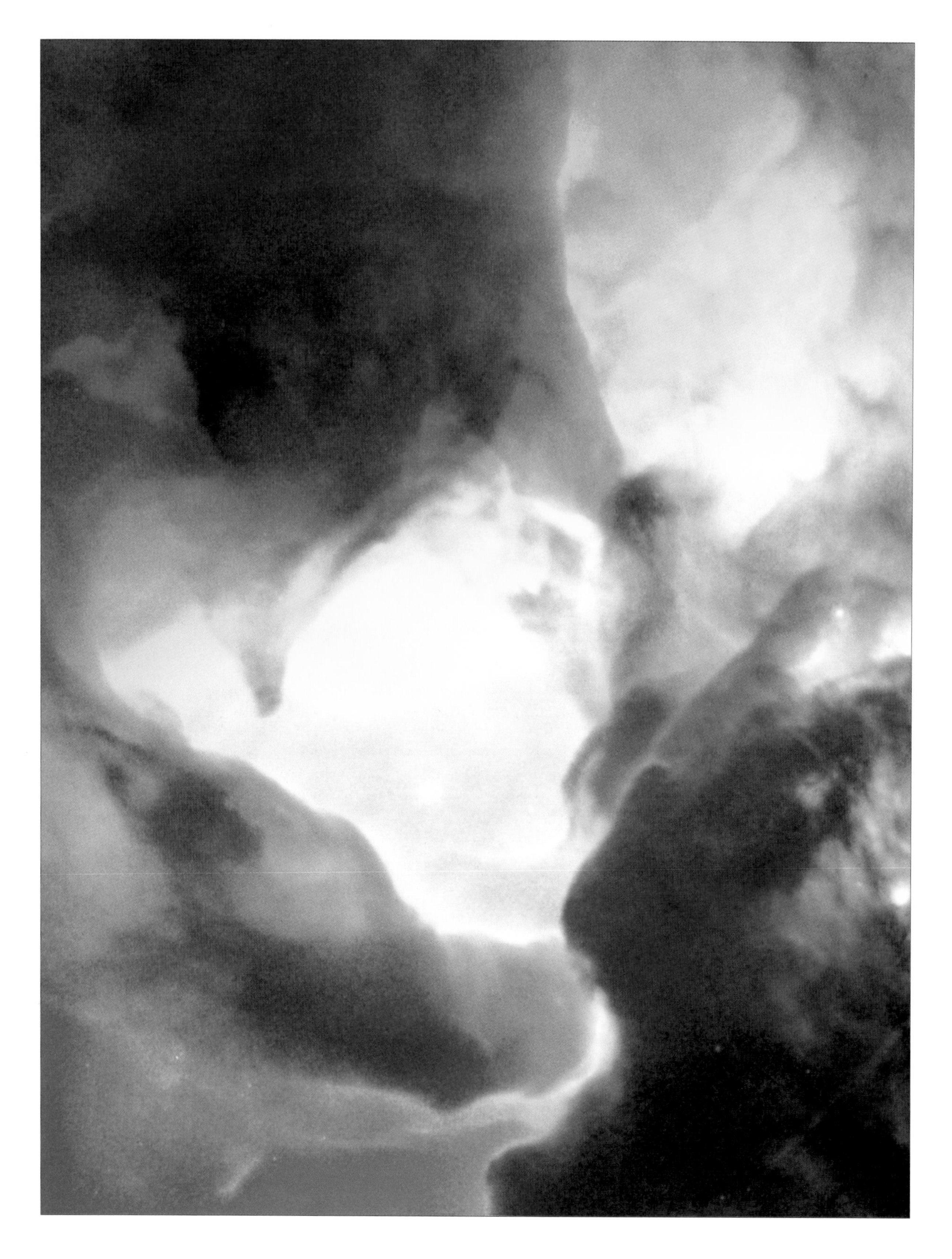

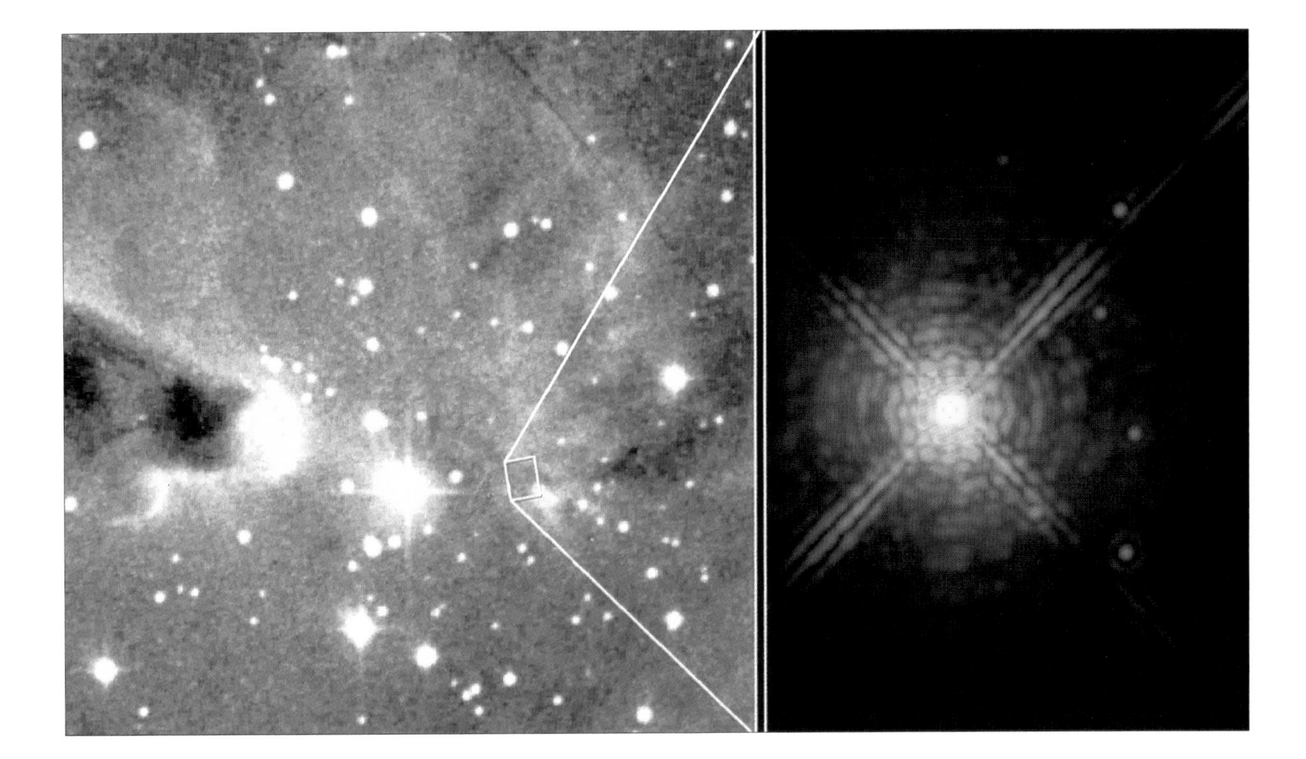

Opposite: This close-up of the region of the Lagoon nebula pictured on the previous page clearly shows the giant 'twister' created by the action of Herschel 36. The strange shape of the twister, which is half a light year in length, is probably produced by the difference in temperature between the heated surface and the cool interior of the cloud formation.

Credit: A. Caulet (ST-ECF, ESA), and NASA

Above: Hubble has revealed a mother star and its six 'baby' stars, hidden within the Cone nebula, 2,500 light-years away from Earth. The image on the left is taken using a ground-based telescope; the Hubble image taken of a detail of the nebula shows the nursery of baby stars which have been formed by material ejected from the mother star during its formation.

Credit: Rodger Thompson, Marcia Rieke and Glenn Schneider (University of Arizona), and NASA

(like a more violent version of the 'wind' of charged particles that streams out from our own Sun and which, when it hits the top of the Earth's atmosphere, is responsible for the Aurora Borealis and Aurora Australis – the Northern and Southern 'Lights'). The wind from the young star hits the surrounding cloud, ionising some of the gas to produce the bright knots.

From its vantage point above the Earth's atmosphere, the Hubble Space Telescope has resolved the mystery of the HH objects in pictures of unprecedented detail. In the final stages of star formation, the cloud from which the star has been born contracts into a disk around the star itself. Material from this 'accretion' disk falls on to the newly ignited star, and while some might be absorbed to form fuel for the future, much is shot out in huge incandescent jets from the poles of the star – out along the axis of stellar rotation. The jets may fire for a brief period – perhaps no more than 100,000 years – but they may also carry away much of the material falling into the star and so limit the fuel available to 'burn' in the future.

Some 450 light-years away, in the constellation of Taurus the Bull, lies HH-30. In its images of HH-30, the Hubble Space Telescope has actually been

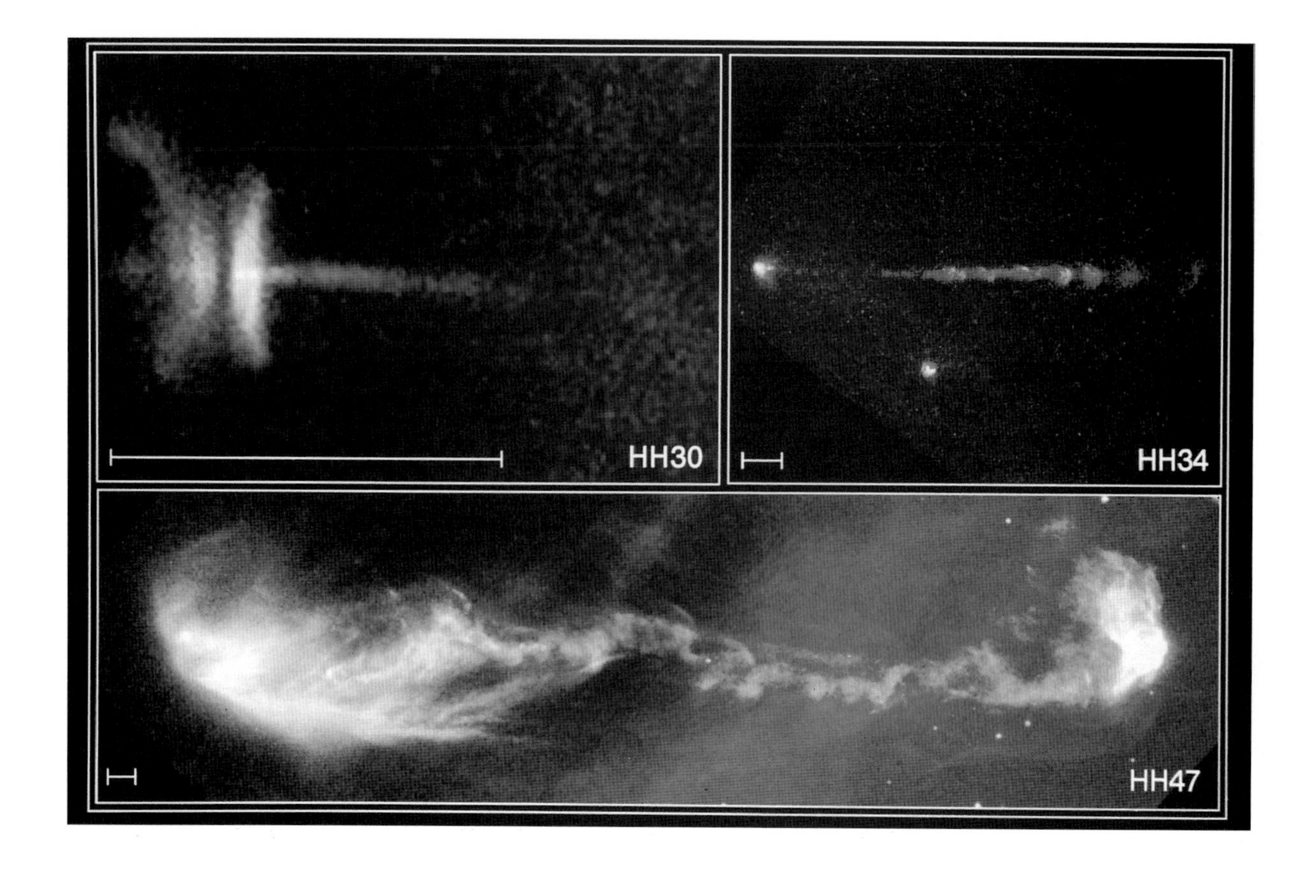

Jets of gas ejected by newly formed stars. Top left: a disk of dust (seen edge-on) encircles a new star, which is producing a red 'exhaust' jet. Top right: a machine-gun-like effect produced as a new star ejects 'bullets' of gas one after the other. Above: a jet 3 trillion miles long, being pushed out by an unseen star at the left edge of the image.

Credits: C. Burrows and J. Morse (STScI), J. Hester (Arizona State University), and NASA

able to make out the disk of material surrounding the new-born star (which itself remains stubbornly invisible behind the densest part). The outer fringes appear to be thicker than those parts closer in, presumably because it takes longer for more distant material to settle into the disk. This is the first time an accretion disk – which in this case is about the size of our own solar system – has been directly imaged surrounding a new star as it forms. The top and bottom surfaces of the disk are lit up by the star and eventually, when the star heats up further and its radiation becomes more intense, it will blow the disk away altogether. (Although, as we saw in the previous chapter, many new-born stars leave enough material for protoplanetary disks and ultimately planets to form.) Material from the accretion disk falling on to the star seems to be responsible for a reddish jet that emerges, presumably direct from the star itself, and stretches out for billions of miles into space. Remarkably, the jet remains tightly confined in a narrow beam over much of this distance. By rephotographing HH-30 a year after the first image, astronomers were able to calculate that the material in the jet is moving at speeds of 500,000 miles (800,000 kilometres) per hour. It is also clear from the Hubble observations that the jet has been emitted in pulses: not so much a continuous jet from a garden hose as a stream of bullets from a machine-gun. This suggests that the material from the circumstellar disk is not falling

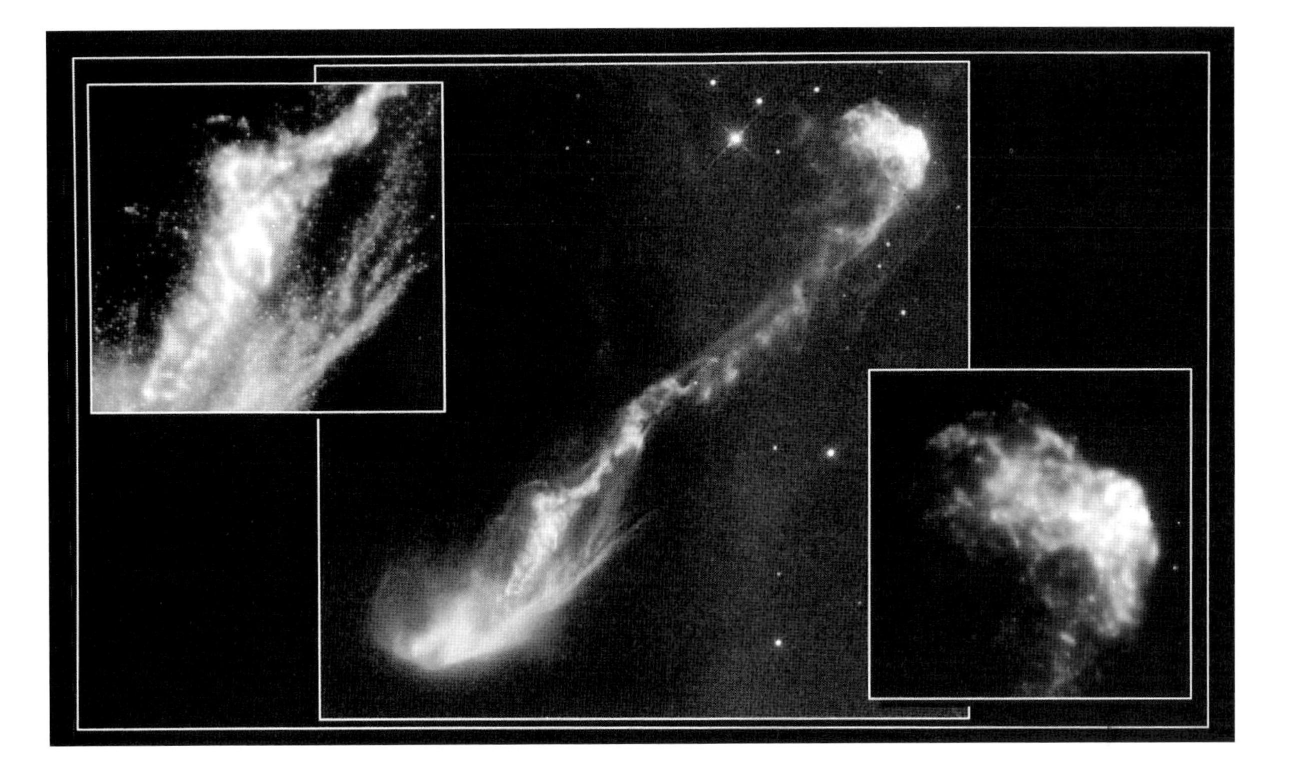

continuously on to the star but occasionally and in large lumps.

Some three times more distant lies HH-34, 1,500 light-years away in the vicinity of the Orion nebula. Here too, despite the great distance, the space telescope has pictured a jet streaming out from a new star. Here again is clear evidence of the machine-gun effect in the formation of these jets. Sometimes the jets are less visible than the blast wave they create as they crash into the surrounding gas. In images of HH1/HH2, also in the Orion nebula, the new-born star remains stubbornly invisible behind a dark, dense, dust cloud, but twin jets (one from the 'north' and one from the 'south' pole of the star) are manifest by their effect in lighting up the clouds. The Hubble Space Telescope can, however, pick up the structure of one of the jets as it emerges from the concealing dust and just before it collides with the dust, revealing once again the chunky nature of the inflow of material to the star.

Another view of the jet produced by HH-47. The enlarged areas show patterns and interactions caused by the repeated collisions between the material shot out in the jet and the gas and dust drifting between the stars. The bottom enlargement demonstrates the bow-shaped shock wave at the jet's leading edge.

Credits: J. Morse (STScI), and NASA

Sometimes, however, the jets are not so pencil-thin and straight. The jet from HH-47, which lies 1,500 light-years away in the southern constellation of Vela, appears to wobble as it is traced out on the sky. The apparent changes in direction of the jet suggest to astronomers that the new-born star that is the jet's source may be one of twins; the star may be wobbling under the influence of the gravitational pull of its companion star, although this has yet to be verified.

But the wobbles are less of a puzzle than the way most of the jets appear to keep straight and narrow for such a long distance. How do they come to be focused into such tight, narrow beams? It may

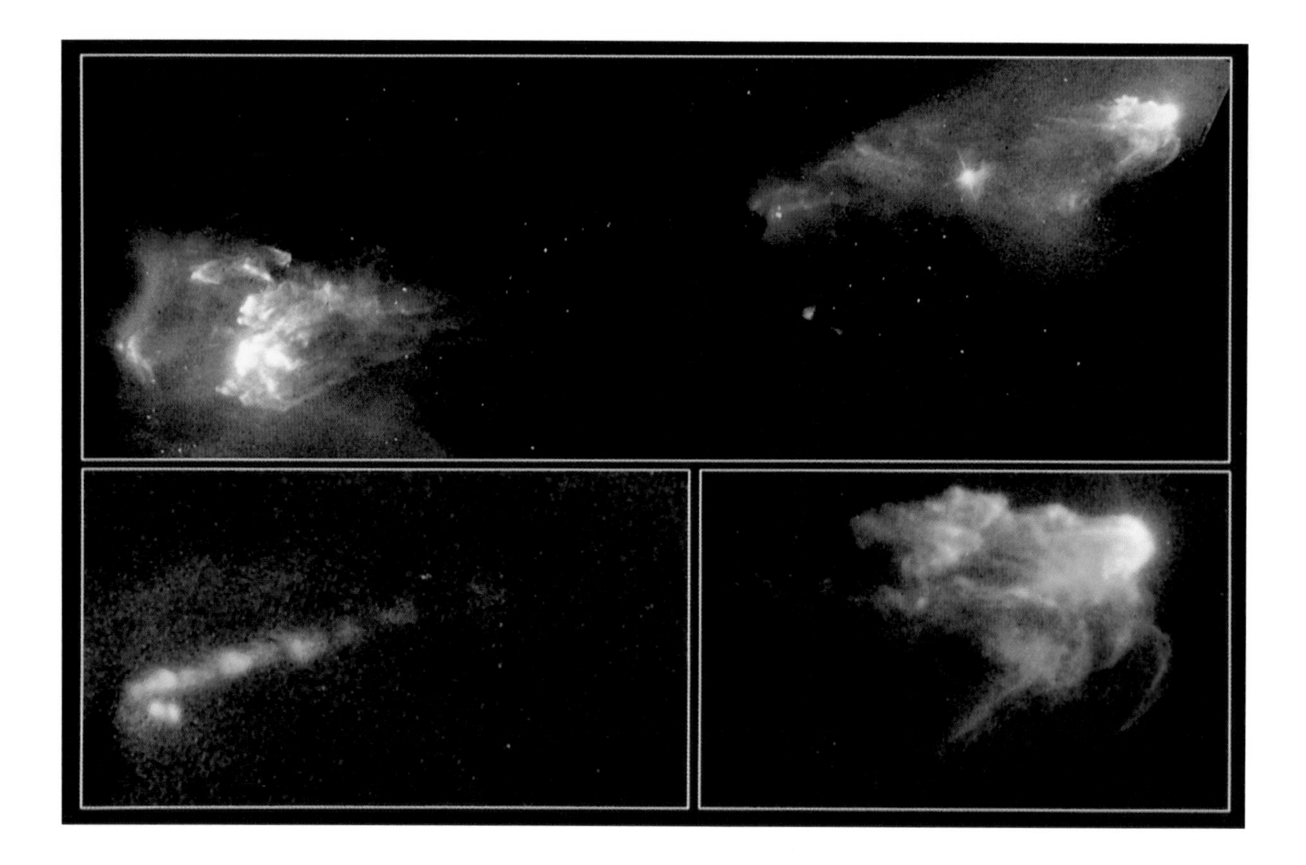

Spectacular views of the double jet (HH1/HH2) being produced by the star hidden in the dark area at the centre of the top image. Note the symmetrical bright areas of gas created at either end of the jet as they slam into areas of slower-moving interstellar gas (the bow-shaped shock wave at the jet's leading edge). The close-ups show (left) the 'bullets' being ejected by the star and (right) the typical bow-wave pattern.

Credits: J. Hester (Arizona State University), and NASA

be that the 'lenses' responsible for concentrating the jets are intense magnetic fields generated by the new young stars, but this too is speculation and more observations are needed before all the details of starbirth are properly interpreted.

Some stars never actually quite make it to the bright lights. To put it in context, our own Sun is a medium- to small-sized star, nothing out of the ordinary in the Milky Way. Stars that are too small – less than 8 per cent of the mass of our own Sun – cannot collapse sufficiently under their own gravity for the temperature and pressure in their interiors to set off a thermonuclear explosion. They never catch alight with the nuclear flame that is the hallmark of the true star. But they do glow in the dark of space, at least for a bit. Their gravitation, although not strong enough to trigger thermonuclear reactions, does cause the star to shrink and it radiates away the energy generated by the motion. These stars are called brown dwarfs.

Astronomers have known for many years that the composition of the universe had to be far more complicated than just big objects (stars) and little objects (planets). They also realised that there was far more mass contained in the universe than could actually be seen, and so formed theories about a range of substellar objects that might exist, could we but see them. Their search for brown dwarf 'wannabe' stars had been fruitless for more than

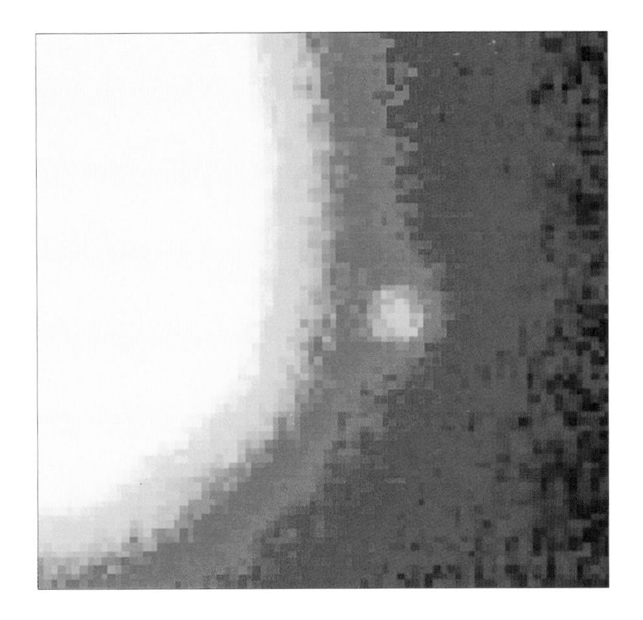

Hubble found the first conclusive evidence of brown dwarfs, which had been long suspected but never clearly observed. The image on the left shows a ground-based telescope's view of a star called Gliese 229, nineteen light-years away from Earth. On the right, Hubble has resolved the star's companion – a brown dwarf called Gliese 229B, which is twenty to fifty times the mass of Jupiter, and is at least 4 billion miles from the star.

Credits: T. Nakajima and S. Kulkarni (CalTech), S. Durrance and D. Golimowski (Johns Hopkins University), and NASA

thirty years, until scientists had use of a space-based telescope. Hubble provided the first unambiguous proof that brown dwarfs actually existed, in a sequence of observations lasting over a year in which scientists working with Hubble co-ordinated their observations with those of ground-based telescopes at Mount Palomar.

The first proven brown dwarf is part of a two-star system, situated in the constellation of Lepus (the Hare) just below Orion; a constellation placed there, according to legend, because Orion liked hunting hares. The brown dwarf is a small companion 4 billion miles (6.5 billion kilometres) distant from a cool red star called Gliese 229, and so is called Gliese 229B. This two 'star' system is a mere nineteen light-years from Earth. (Nearby stars are referred to by Gliese numbers because in 1969 the German astronomer W. Gliese published the most comprehensive catalogue of stars within eighty light-years of Earth.) Gliese 229B is 100,000 times dimmer than our own Sun, and as such is the faintest object to be spotted to date orbiting another star. Gliese 229B is too large and too hot to be classified as a planet (it has between twenty to fifty times more mass than our solar system's largest planet, Jupiter), but is nevertheless just too small and cool to shine like a star.

Such 'failed' stars represent a missing link between planets and stars, part of what has been termed the 'twilight zone' of substellar objects. The analysis of Gliese 229B will help astronomers distinguish between massive Jupiter-like planets and brown dwarfs orbiting other stars, thus assisting in the search for planetary systems beyond our own solar system. Without being able to make such a distinction, many brown dwarfs orbiting other stars might have been mistaken for planets; for example, Gliese 229B is so relatively cool that methane, a compound typical of gas giant planets but not of stars, has been detected in its chemical composition.

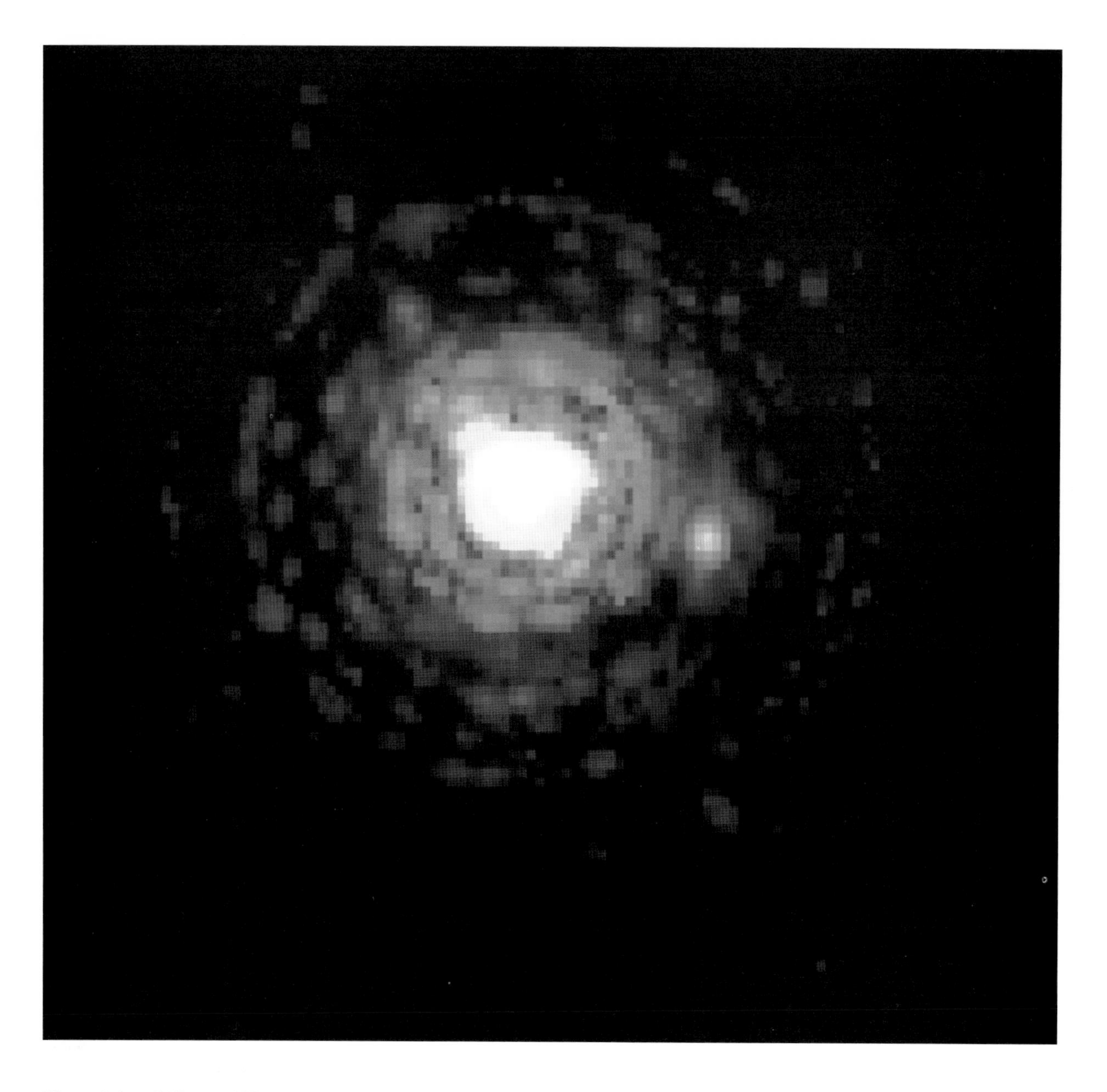

The red dwarf Gliese 623B, one of the smallest stars in our galaxy, which is to the right of centre of the picture. The red dwarf is part of a double star system, where the two stars are separated by little more than twice the distance between the Earth and the Sun (i.e. about 200 million miles). Astronomers had long suspected the red dwarf's presence from the fact that the main star 'wobbles', due to the gravitational influence of a nearby object – which remained unseen until it was revealed by Hubble.

Credit: C. Barbieri (Univ. of Padua), and NASA/ESA

Brown dwarfs still store enough internal heat from when they first collapsed to glow strongly in the infra-red. In the case of Gliese 229B, at least twenty times the mass of Jupiter is squashed into a space the size of just one Jupiter, so the amount of heat stored from its formation is considerable, and the dwarf's temperature is between 500 and 700 degrees Celsius. But the future for Gliese 229B, like all brown dwarfs, is quite literally not a bright one. Deprived of the process of nuclear fusion, they have no internal source of heat to replenish what they radiate away. One of the reasons astronomers have searched fruitlessly for them over three decades is that as time goes on they become cooler, fainter and ever more difficult to see. Eventually, they just quietly fade away into the darkness.

Other stars, like Gliese 229 itself, just make it over the threshold and ignite to 'burn' hydrogen in a nuclear inferno at their core. Stars that are heavier than brown dwarfs but lighter than our own Sun tend to be cool, small and red. They too are difficult to observe. These red dwarfs were once thought to be among the most abundant stars in the Milky Way, but observations by Hubble have shown that this is not so. In June 1994 the telescope pictured a red dwarf for the first time, one of the smallest true stars in our galaxy, about twenty-five light-years away in the constellation of Hercules. Again a nearby star (the more distant ones are simply too faint to be observed, even by Hubble), Gliese 623B's existence had been predicted because its companion star had 'wobbled' in a way that suggested it was being gravitationally tugged by a small star very close by. But ground-based telescopes had been unable to spot the red dwarf, which is only one-tenth the mass of the Sun and about 60,000 times less bright, and so was outshone by its larger companion star. The Hubble Space Telescope was able to tell the two stars apart for the first time – to 'resolve' the red dwarf from its companion.

Red dwarfs are small and cool; their surface temperatures are around 3,200 degrees Celsius, whereas the surface of the Sun is around 5,500 degrees Celsius. But as if to compensate, red dwarfs go though their fuel very slowly. If it is typical, Gliese 623B will still be shining in 100 billion years' time – provided the entire universe lasts that long – which in stellar terms will make it very long in the tooth indeed.

By that time, our own Sun – and therefore life on Earth – will be long dead. The Sun is a perfectly ordinary, middle-aged star. Because it is hotter than a red dwarf, it appears yellow in colour, but also because it is hotter it will run through its fuel more quickly, perhaps in 5 billion years' time. There are some stars that are still hotter and that shine white or even blue, with a surface temperature ranging up to 25,000 degrees Celsius, almost four times as hot as the surface of the Sun. Such stars are believed to be about fifty times as massive as the Sun, but are millions of times brighter.

Such bright stars are rapidly snuffed out. For example, a star which is ten times as massive as the Sun will burn its fuel about 5,000 times faster. The gravitational force of all the extra material squeezes the core of the star more tightly, raising the temperature and pressure in the interior and speeding up the fusion reactions. Such a star will burn itself out in just 20 million years – in cosmic terms, the blink of an eyelid.

A star's colour and its temperature depend on how massive it is and on the traces of material other than hydrogen of which it is composed. All these things, including its mass, depend on the material from which it was made: the composition and the quantity of gas and dust in the clouds and Bok globules from which the star was born. In other words, a star's future and fate is determined from the moment of its conception. It is a neat irony that while modern astronomy has proved astrology false – disproving the notion that the position of the stars at a human's birth affects his or her subsequent fate – it has demonstrated that the lifetime, the lifestyle, and the manner of a star's dying are all written in the circumstances of its birth. The conditions of a star's birth do indeed determine its ultimate fate. And it is to the many deaths of the stars that we turn in the next chapter.

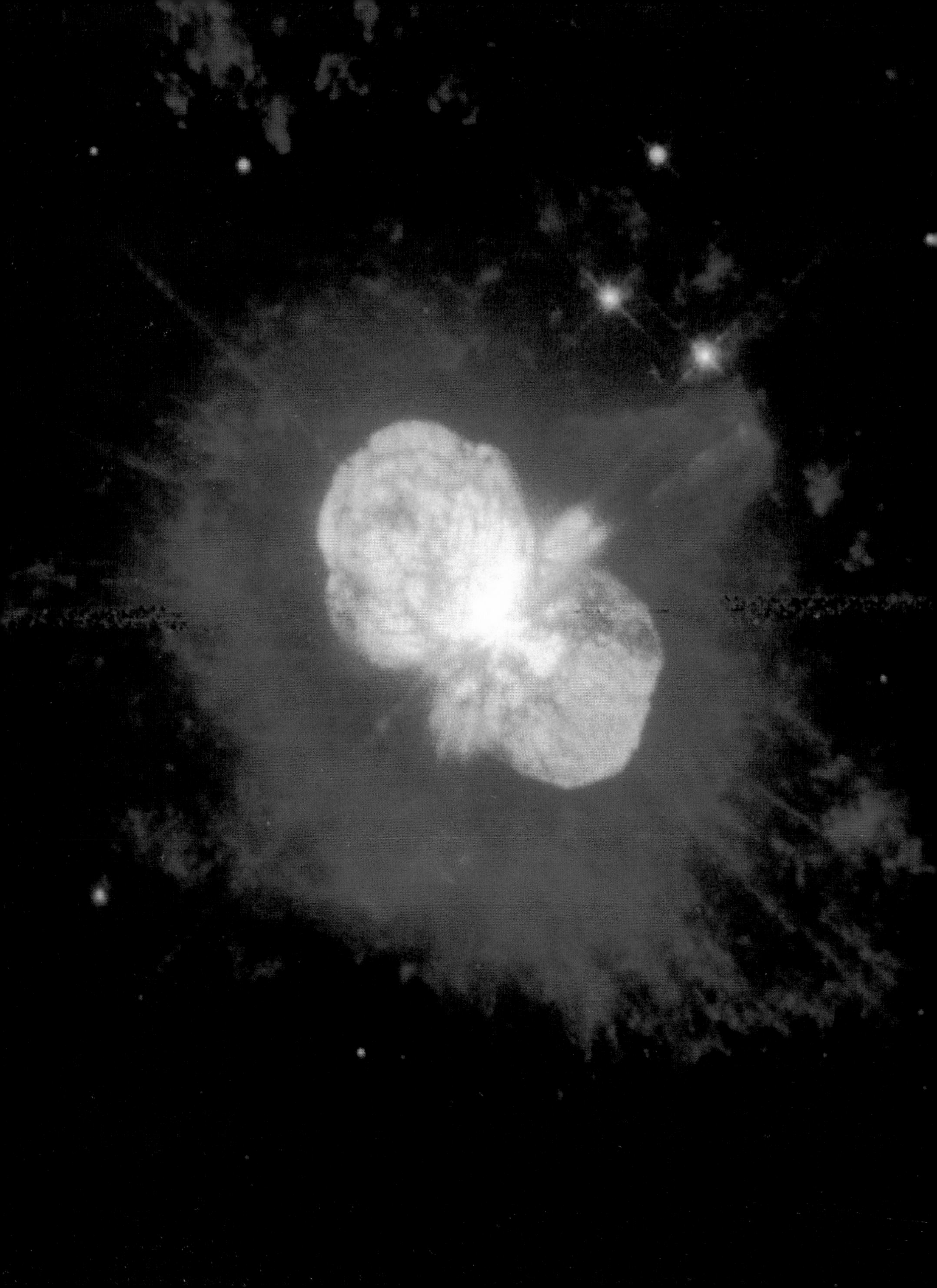

CHAPTER FOUR

INTO THE FURNACE

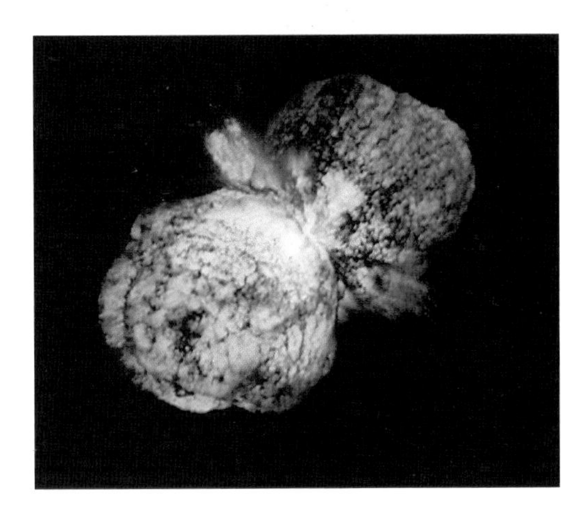

Even the stars die. Many deaths await them. There is the star that rages against the dying of its own light, and in a last burst of thermonuclear power swells up and blasts its outer layers into space in a supernova explosion. The enormous forces involved transmute its hydrogen and helium into the other chemical elements. It is in these colossal furnaces that gold and silver and the other elements familiar to us here on Earth are made and then scattered into interstellar space. Aeons later, they will collect together in orbit round another, younger, star and come together to form rocky planets.

The supermassive star Eta Carinae pictured just after the new Wide Field and Planetary Camera 2 had been fitted to the Hubble Space Telescope to compensate for the defect in the main mirror (*opposite*), and again two years later (*above*). Possibly the most massive star known in our galaxy, more than 100 times as massive as the Sun, Eta Carinae has been puzzling astronomers for more than 150 years.

In 1833, it started producing a great burst of energy and by 1841 had become the second brightest star in the sky, radiating more than four million times as much energy as our Sun. The explosion produced this pair of huge, billowing clouds of gas and dust — two polar lobes — and a large thin equatorial disk. The material is all moving outward at speeds of between 1.5 and 2 million miles an hour.

These natural colour pictures are a combination of three different images taken in red, green, and blue light. The reddish outer glow comes from the fastest moving material ejected during the last century's outburst. It is largely composed of nitrogen and other elements formed inside the star.

Credits: J. Hester (Arizona State University) and NASA (above)
J. Morse (University of Colorado) and NASA (opposite)

That was how the Earth – and humanity – was formed. The hippies of the 1960s were literally correct: we are all stardust.

Because of their great distance from us, the light from these cosmic cataclysms is attenuated and faint by the time it reaches us. But some supernova explosions have been so massive that they have shone out in the sky for all on Earth to see.

While they are spectacular, supernova explosions are by no means the general fate of the stars. Some expire quickly and ignominiously, going quietly into the endless night. As we saw in the previous chapter, some 'stars' do not even manage to catch alight with nuclear fire, but shine with the energy produced by their own contraction. As their heat radiates away, and with no inner power source to replenish the energy, they quickly cool and darken. The ultimate fate of these brown dwarfs is to fade into black dwarfs. Other stars modestly husband their resources of fuel and seem to last for ever. These are the red dwarf stars, less massive than our own Sun, but whose interiors none the less reach high enough temperatures and pressures for nuclear fusion. They carry on and on, shining for 100 billion years or more.

Perhaps of greatest concern to humanity is the fate of our own local star, the Sun, for that will determine in turn the fate of the Earth. In contrast to the immense longevity of the red dwarfs, the Sun will spend 'only' around 10 billion years burning hydrogen, and has about 5 billion years to go. Currently, the pressure at the core of the Sun – the innermost quarter of the star – is about 100 billion times the atmospheric pressure at sea level on Earth, and the temperature of the core is around 15 million degrees Celsius. These conditions are hot and dense enough for nuclear fusion. In the core of the Sun at this very moment, something like 600 million tonnes of hydrogen are transmuted into roughly 596 million tonnes of helium every second. The 'missing' 4 million tonnes of material are converted straight into energy in the form of radiation in accordance with Einstein's famous equation $E=mc^2$ (energy equals mass times the speed of light squared). This vast energy output generates huge convection currents in the Sun's outer atmosphere as heated gas wells up from the interior. At the surface the temperature is about 5,500 degrees Celsius, and the energy radiates out into space in the form of visible light, X-rays, ultraviolet, infrared and radio waves.

It is when the fuel finally runs out that things become uncomfortably interesting in the neighbourhood of a star. Stars remain stable during their lives because the fantastic energy produced in their internal furnaces generates an outward pressure that counterbalances the equally fantastic weight of their own outer layers. But the apparent stability is actually a delicate balancing act. When the fuel runs out and the intense source of energy in its centre starts to shut off, the Sun and stars like it collapse under their own colossal gravity. This heats them up again until the shell of hydrogen outside the (helium-filled) core ignites in nuclear fusion. This causes the star to swell gradually into a type of star known as a red giant. Sad to relate, at this stage the Sun will start to consume its own: the inner planets, including Earth, will be engulfed as the outer layers of the Sun bulge outwards, extending perhaps beyond the orbit even of Mars.

Humanity may very well not be around to witness the beginning of its own end. We may by then have blown ourselves up through an excess of cleverness, or lost the planet to a more dominant species (insects seem the likeliest candidates). Or we may have moved house, using interstellar travel to colonise new planets in other, younger systems. But if our most distant descendants are still living on Earth 5 billion years hence, they will face a ghastly and fiery fate.

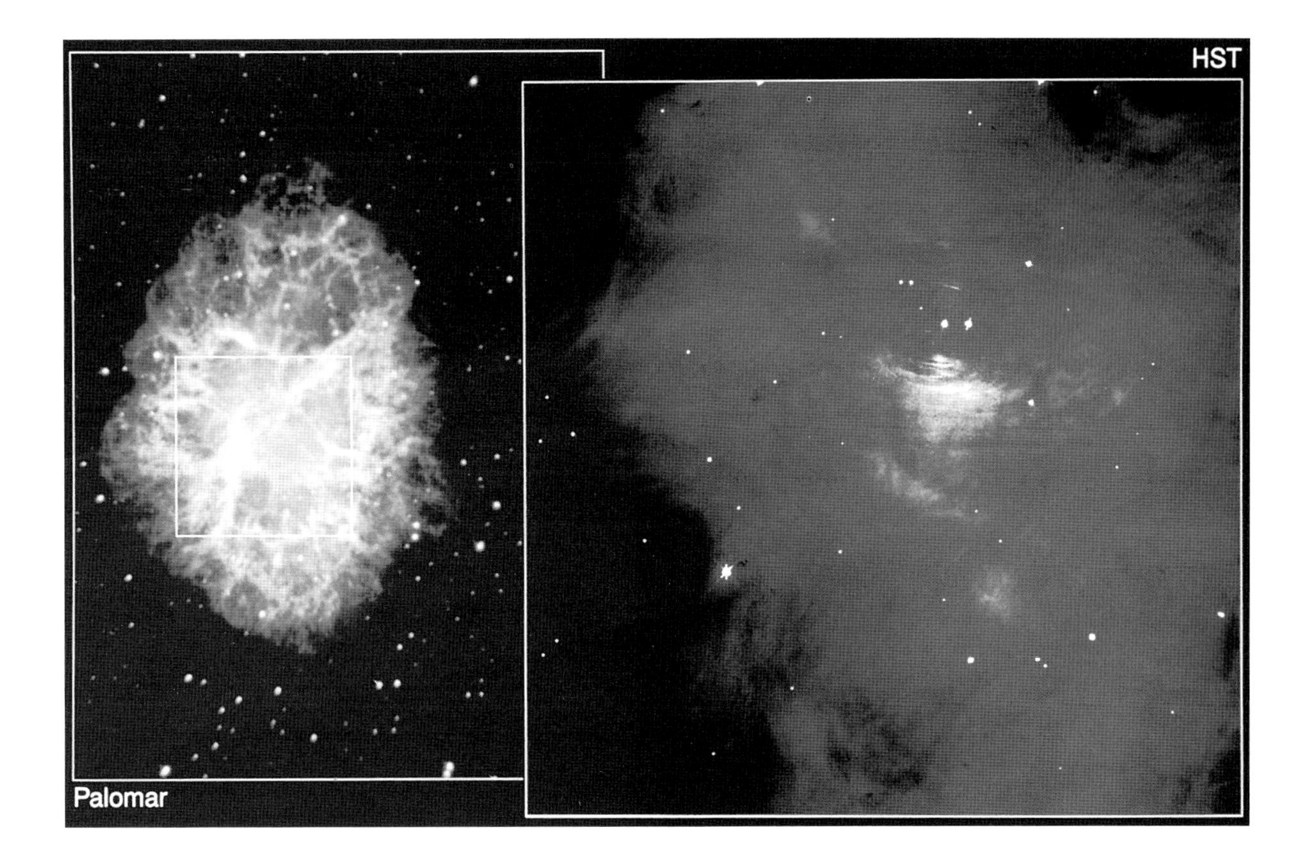

The Crab nebula is all that remains of a massive stellar supernova explosion which was observed by Chinese astronomers in 1054. At its heart lies a pulsar 6 miles across, the tiny core of the shattered star.

Credit: J. Hester (Arizona State University), and NASA

As the ageing Sun swells, a red monster filling half of the sky, Earth will wilt under the terrible heat; the atmosphere will be blasted away, and Earth will become an arid, airless, barren rock – much like Mercury today. Relentlessly, the heat will increase until the rocks start to glow, incandescent. Then the flaming outermost layers of the gross and bloated object that our star has become will begin to lick at the surface of the Earth. The rocks will vaporise and vanish. The Earth will be no more; not even a cinder will remain. Our planet and all that it contains will be stripped to its elemental atoms, which will be blown into interstellar space. Perhaps, billions of years later, some of these atoms may find themselves coalescing in a giant molecular cloud to be recycled in a new star and a new planetary system.

(One should not get too depressed at the thought of this end, for it lies almost unimaginably far into the future. The Earth has been in existence for some 4.5 billion years. There is some evidence of microscopic life – bacteria and the like – in rocks that are about 3.5 billion years old. Complex life forms evolved only at around the time of the Cambrian period, roughly 530 million years ago. Anatomically modern humans are only some 150,000 years old. Thus the Earth will remain untouched for about 30,000 times as long as humans have already been in existence.)

With the planets reduced to dust flitting through space, the Sun's death throes will continue. As the hydrogen in the outer layers fuses, the helium produced will sink back to the still-shrinking

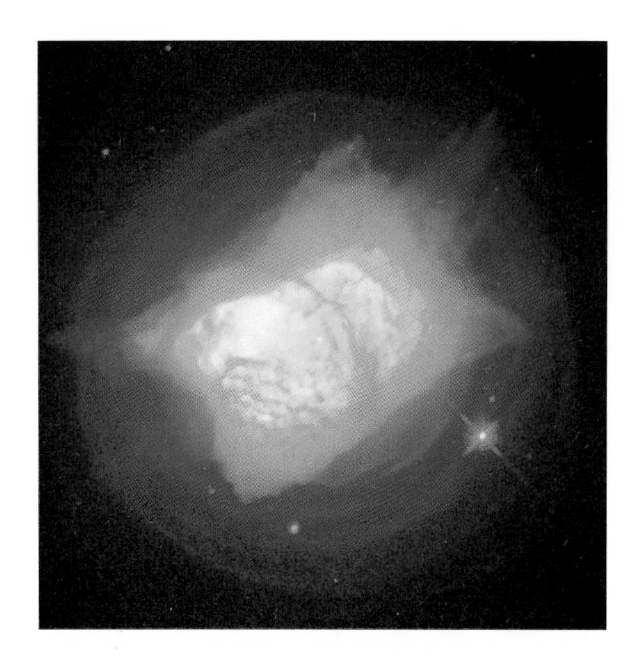

Planetary nebula NGC 7027, some 3,000 light-years distant, in the direction of Cygnus the Swan. This false-colour superposition of two Hubble images shows that at the red giant stage, the star ejected material from time to time and in a symmetrical fashion to produce the blue concentric shells. Later, all the remaining outer layers were blasted off in an irregular fashion and the bright dense clouds of dust condensed from the ejected material.

Credit: H. Bond (STScI) and NASA

core, which will become steadily more and more dense. Eventually, when the temperature at the core reaches about 100 million degrees Celsius (i.e. ten times the current temperature), a crisis point is reached. This is called the 'helium flash', when the helium atoms themselves begin to fuse together and produce carbon atoms together with some oxygen. Like a build-up of ashes or soot in a conventional fire, the carbon will choke off the nuclear reaction in the core, which will collapse as a result. The heat generated by the squeezing of the core will rise up to the outer layers which, in one last thermonuclear gasp, will blow out the remaining material. These ejected layers – which sometimes look like smoke rings in astronomers' telescopes – have been given the confusing and inaccurate name of planetary nebulae. (They were called such by the great Anglo-German astronomer William Herschel who, over a period of twenty years, studied more than 2,500 nebulae of all sorts of types and origins. He coined the name for the planetary nebulae in 1785 because at first these clouds of gas looked both nebulous and yet had a disk like a planet – resembling distant planets like Uranus.)

The material blown out from the outer atmosphere will travel away as an expanding ring of gas, glowing as a result of the energy it receives from the remnant left behind. All that will remain of the Sun will be a very small, very hot star, known as a white dwarf. It will be about the same size as the Earth used to be before it was blasted into smithereens, but 10,000 times more dense. Ordinary matter – wood, stone, water – is largely empty space. It is composed of atoms, with a tiny nucleus at the centre of each atom surrounded by even tinier electrons which orbit the nucleus a long, long way away. By contrast, the enormous pressure inside a white dwarf crushes matter to a 'degenerate' state where the electrons are crammed much closer together. Degenerate matter is so dense that a teaspoonful of material from a white dwarf star would weigh more than a tonne.

The surface temperature of a white dwarf is very high: around 17,000 degrees Celsius, nearly three times the temperature of the surface of the Sun today. However, inside a white dwarf it is still too cool for thermonuclear fusion to take place: the interior is around 1,000,000 degrees Celsius, less than a tenth of the temperature at the core of the Sun today. With no inner powerhouse to replenish supplies of energy, white dwarfs too suffer the ignominious fate of slowly dimming over billions of years until all that is left is a dark cinder of degenerate matter – a black dwarf, burned out and floating invisibly in the darkness of space.

The Hubble Space Telescope has allowed us a preview of the future fate of the Earth and the Sun. Astronomers have used it to observe and analyse light from many other stars going through the stages of ageing, swelling and dissolution that await our own star. They have also watched many other types of star dying. Apart from providing

Some 3,000 light-years from Earth, the central star of the Egg Nebula (also known as CRL2688) was a red giant only a few hundred years ago. The nebula itself is the large cloud of dust and gas ejected by the star. This image (shown in false colour) was taken in red light with the Wide Field and Planetary Camera 2 and clearly shows bright arcs cutting across beams emanating from the hidden central star.

Credit: R. Sahai, J. Trauger (JPL), the WFPC2 science team, and NASA

spectacular images of celestial fireworks, the death processes of stars hold the keys to understanding the formation and birth of other cosmic objects. As they die, the stars hurl out carbon and many other elements that will provide the building blocks of new stars and planets.

One thing that has surprised astronomers is the great variety of shapes assumed by planetary nebulae. Once it had been thought that most red giants, being basically round, would produce neat, spherical clouds of gas when they exploded, which would simply fly out into space, getting larger and more tenuous as time passed. But the pictures taken by Hubble have shown clouds of gas resembling the streams thrown out by lawn sprinklers, Catherine wheels, rocket-engine exhausts: all sorts of different shapes, some highly symmetrical and some all over the place. They have also seen strange dough-nut-shaped disks of dust surrounding dying stars, balloons of gases expanding inside larger balloons, red blobs scattered along the edges of nebulae . . . Scientists suspect that the intriguing shapes may be created by hidden hands – by the action of unseen companions to the dying stars, whether brown dwarfs, smaller stars, or even large planets. Bruce Balick, of the University of Washington in Seattle, whose team obtained some of these breathtaking pictures, commented: 'We're going to have to confront these strikingly symmetric structures with some fundamentally revised ideas about the final stages of a star's life. The lovely patterns of gas argue that some highly ordered and powerful process orchestrates the ways stars lose their mass, completely unlike an explosion.'

One of the most fascinating dying stars is Mira, a star that has been known and observed for nearly 400 years but which retained some of its secrets for all that time until Hubble provided the first ultraviolet images of the red giant and its nearby small hot companion. Getting images that managed to separate the two stars was quite an achievement; the system is so distant that when seen from Earth the two stars take up as much of the sky as a coin placed four miles away. Mira is a variable star: its brightness fluctuates over a period lasting 332 days. It got its name from its discoverer, the Dutch clergyman David Fabricius, who first saw the star in 1596. He named it Mira, meaning 'The Wonderful'. (Sadly, Fabricius's career as an astronomer was cut short by his murder, allegedly at the hands of one of his parishioners suspected of stealing the pastor's goose.)

Once like our Sun (it has approximately the same mass) Mira is now at the end of its life. It has swollen into a cool, red giant which regularly contracts and expands, creating a powerful solar 'wind' that pushes vast amounts of dust and gas into space. Hubble has also taken pictures in visible light which shows that Mira has an odd shape, not spherical but more like a rugby ball. The Space Telescope has allowed astronomers to measure the size of Mira, despite the fact that it is 400 light-years – more than 2.4 million billion miles (4 million billion kilometres) – away. They discovered that Mira is 700 times the diameter of our own Sun. If our star goes the same way as Mira, when it reaches the red giant phase it will become so large that its surface will reach almost two-thirds of the way to Jupiter.

Mira's companion star is a white dwarf. The ultraviolet images show a hook-like spur stretching out from Mira towards its companion, which could be material being drawn away by gravity. The white dwarf is surrounded by material captured from Mira's solar wind.

(Our Sun is unusual in being a singleton. Most stars are (non-identical) twins, waltzing around each other for their entire lives, bound together by the attraction of their mutual gravitation. A few appear to be in systems consisting of three or even four stars.)

Mira is still in the preliminary stages of jettisoning the outer layers of its atmosphere.

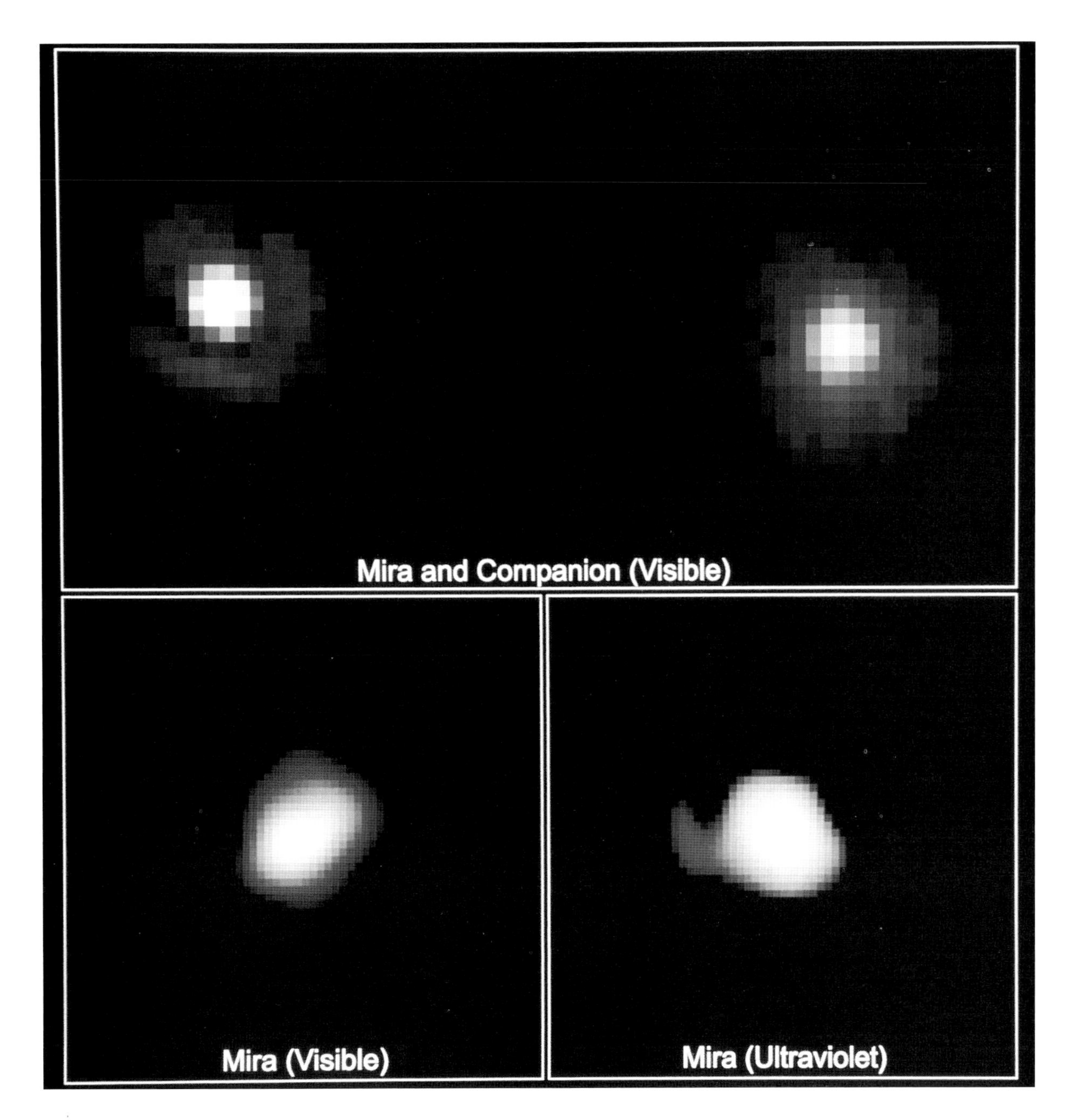

The double star system of Mira (Omicron Ceti) in the constellation Cetus. The main star (top right) is about as massive as our Sun but, now towards the end of its life, has swollen into a red giant roughly 700 times the Sun's diameter. Mira is a variable star. Its companion is a burned-out star, a white dwarf, surrounded by material captured from Mira's 'wind' of gas and dust. The UV image shows a small 'hook' coming from Mira, in the direction of the companion.

Credit: M. Karovska (Harvard-Smithsonian CfA) and NASA

Hubble has pictured later stages of the death throes of stars like our Sun, and has shown that it is a complicated process which astronomers cannot yet explain in all its detail. Much information has been drawn from Hubble's study of planetary nebulae, which (in astronomical terms) are created very rapidly indeed, i.e. over a few tens of thousands of years.

One Hubble image is of a planetary nebula seen relatively early in the process. It is called NGC 7027 (meaning that it is object number 7027 in the New General Catalogue of clusters and nebulae first compiled by the Danish astronomer J.L.E. Dreyer in 1888; matters had advanced since Herschel's study of 2,500 such objects a century earlier). Lying 3,000 light-years away in the direction of the constellation Cygnus, the nebula resembles an onion in the way layers of material are piled one behind the other. The Hubble image clearly shows how faint blue concentric shells surround the nebula, and the bright inner region is shot through with an extensive network of red dust clouds. The white dwarf itself is visible as a white dot in the centre. It is due to the radiation from the white dwarf that the clouds of dust and gas can be seen at all, because the radiation heats the ejected material – to the point where the atoms are stripped of their outer electrons in a process known as ionisation – so that it glows in the dark. (The picture, by the way, is not in true colour because it is actually a composite of two images, one taken in visible light and one in infra-red.)

The Hubble image indicates that the star did not blow off all its outer atmosphere at once but puffed it out in stages, thus forming the concentric shells. In a final gasp, the remaining outer layers were spat out to produce the bright inner regions seen in the picture. This process was disordered and not symmetrical, and dense clouds of dust have subsequently condensed.

Even more striking are the images of the Egg nebula, which again is an early stage in the death throes of a red giant that was once about the same size as our Sun. It lies some 3,000 light-years distant. What look like the beams of searchlights emerge from the central star, which itself lies hidden in a dense band of dust, visible as the dark line across the centre. The 'beams' appear to punch through concentric bright shells of material. These shells are dust and gas blown off the star and expanding at a rate of 20 kilometres a second or nearly 45,000 miles per hour. The patterning of bright bands with darker rings in between suggests that the material has been blown off the star in short bursts of perhaps 100 years, followed by a pause, then another burst, and so on.

But the details are perplexing. Are the 'searchlights' really starlight escaping from holes in the band of dust? If so, how were these holes formed? Perhaps, scientists speculate, they were carved out by an escaping jet of matter streaming out from the central star. But the mechanism for producing such a jet of material is not understood; presumably the star at the centre was originally spherical like our own Sun, so how did it give rise to this highly unspherical pattern?

Hubble has had a look at the Egg nebula not only in visible light but also in infra-red, which can pick up details invisible to the human eye. The infra-red picture shows starlight reflected off dust particles (shown in blue) whereas the red areas indicate hot molecular hydrogen gas. The picture clearly shows how the high-speed ejection of a stream of material is hitting the slower outflowing shells, with the heat of the impact causing the hydrogen molecules to glow.

Yet more questions have been thrown up by Hubble's study of MyCn18, another planetary

Opposite: This strange and unsettling image which resembles an unblinking eye staring out of the darkness of space is the 'hour-glass' nebula, MyCn18, a young planetary nebula situated about 8,000 light-years away. The picture is a superposition of three separate images showing the light emitted by ionised nitrogen gas (red), hydrogen (green), and oxygen (blue). The hour-glass shape results from the expansion of a fast stellar wind inflating a slowly moving cloud which is denser at the equator than the poles.

Credit: R. Sahai, J. Trauger (JPL), the WFPC2 science team, and NASA

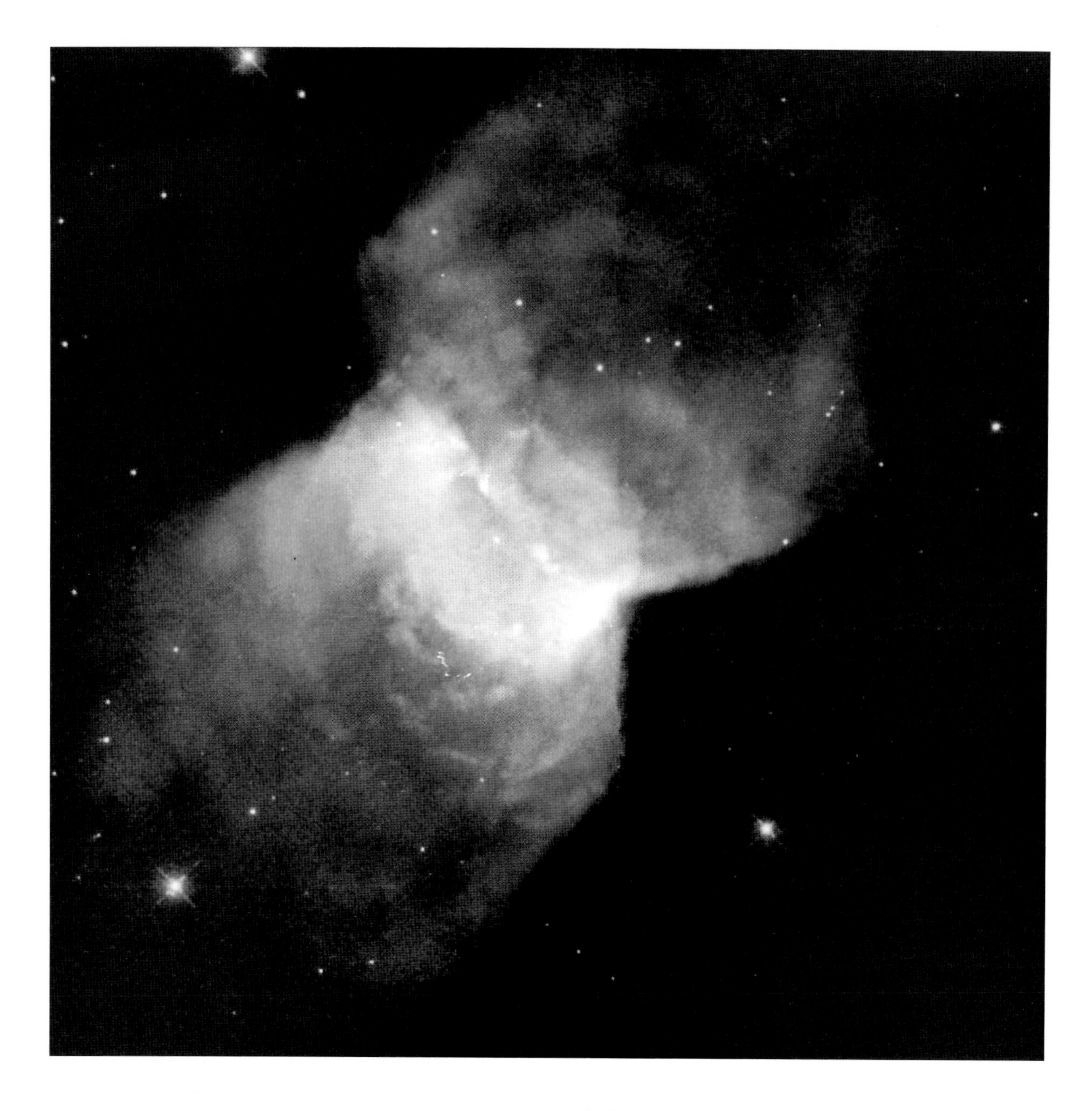

Pretty, shaped like a butterfly wing, yet the nebula NGC 2346 is the result of stellar cannibalism. The stars at the centre were so close that when one expanded to become a red giant, it swallowed its companion. The two spiralled in together, and the outer layers of the red giant were blasted off into a disk, which can be seen in the image. A fast stellar wind followed, puffing out the surrounding disk into the large, wispy wings.

Credit: M. Stiavelli (STScI), NASA, and I. Heyer (STScI)

nebula about 8,000 light-years away. When viewed by ground-based telescopes, MyCn18 had seemed rather simple: a pair of large outer rings with a smaller, central one. The Hubble Space Telescope picture tells a different story. MyCn18 is a complicated hour-glass shape with various arcs or shells of material one behind the other; perhaps the remains of discrete layers of material ejected by the dying star, although there may be other explanations. The picture is in fact a composite of three images taken to show light from ionised nitrogen (which shows up as the red rings), hydrogen (green) and oxygen (blue). It highlights a central puzzle for astronomers interested in the evolution of the stars: how did a star that presumably was round like our Sun give rise to such a complicated pattern of rings?

One theory for the hour-glass shape is that it is produced by a fast solar wind – a stream of energetic particles pouring out from the central core of the star – expanding rapidly within a much more slowly moving cloud of material. If the material is denser near the equator rather than the poles, then an hour-glass shape might result. But the central region – whose details were not visible to ground-based telescopes – appears to make this neat explanation rather difficult to stick. For it is not a dense region near the equator at all, but a complex structure with its own pair of intersecting elliptical rings: an hour-glass within an hour-glass. And to cap it all, the hot star that is supposed to be the source of all this material, and from which the radiation streams out to illuminate the nebula, is not at the centre, but is offset. As so often, Hubble's new information has set astronomers fresh mysteries with which to wrestle.

Puzzles abound even closer to home, in the Helix nebula. This is the closest planetary nebula to Earth, just 450 light-years away in the constellation of Aquarius. Seen through ground-based telescopes, this looks like a classic, apparently symmetrical ring-shaped nebula. Although one might, fancifully, see it as the pupil of a giant eye hanging in space looking down on us with an unblinking stare, it can equally and more accurately be interpreted as a near-perfect smoke ring, the shell of gas and dust blown off from a dying star. But seen through the Hubble Space Telescope, intriguing details appear which are taxing astronomers' ability to explain the precise dynamics of the nebula's formation.

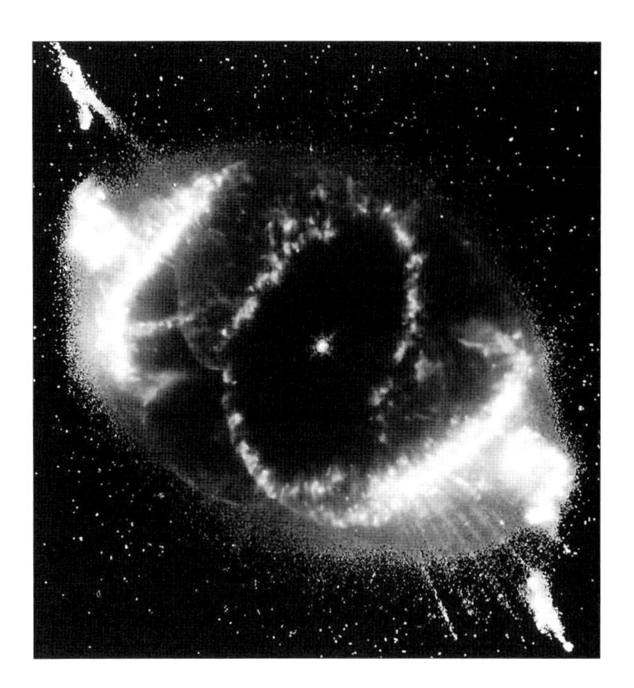

The richly complicated dynamics of the Cats's Eye nebula (known officially as NGC 6543) are revealed by this composite of three images taken at different wavelengths by the Wide Field and Planetary Camera. At the stage of stellar death shown in this image, a tenuous, fast wind is blowing from the star to sweep up the material already ejected to form the nebula. The fast wind is thus creating an inner bubble.

Credit: J.P. Harrington and K.J. Borkowski (University of Maryland), and NASA

The Hubble pictures show thousands of what look like tadpoles or sperm wriggling in space around the dying star. Robert O'Dell (who examined the protoplanetary disks in the Orion nebula described in Chapter Two) first discovered the objects and called them 'cometary knots' because, with their head and tail structure, they resemble the comets seen in our night skies. Appearances can be deceptive, for these objects are huge. The head of each 'comet' is in fact twice the size of our entire solar system, and the tails stretch for billions of miles. They are far too large to be true comets; they seem to be a consequence of the way in which successive outer layers of the dying star were blasted off one after the other. Towards the end of the

Above and opposite: The Helix nebula, NGC 7293, is the closest planetary nebula to Earth, just 450 light-years away in the constellation Aquarius. The HST has been able not only to provide a close up of part of the ring but also to yield information on the dynamics of the explosion of the 'smoke ring' blasted off from the dying star in the centre. By looking at the light emitted from different gases (nitrogen is shown in red; oxygen in blue and hydrogen in green), the image reveals what happens when streams of gas collide. The objects which look like tadpoles have heads each of which is twice the size of our solar system while each tail stretches about 1,000 times the Earth's distance to the Sun. The gaseous knots are the results of a two-stage ejection of material from the dying star. Hot gas rejected from the star collides with cooler gas that had been thrown out 10,000 years earlier. The shock fragments the smooth cloud surrounding the star into droplets.

Credit: R. O'Dell, K.P. Handron (Rice University), and NASA

process, a faster-moving shell of hot gas will catch up and collide with slower-moving gas released some 10,000 or so years earlier. The hot, low density gas mixes with the cooler and denser older shell and fragments the cloud of gas, breaking it up into smaller and denser 'droplets'.

The cometary knots appear to lie along the inner edge of the ring, with their tails stretching outwards and their heads pointing towards the interior; rather as the tails of comets stretch away from the Sun as the material from the cometary head is boiled off by heat and radiation, so presumably the gossamer tails of the knots are created in a similar way, blown by the intensity of the radiation from the dying star at the centre of the nebula.

In a few hundred thousand years, the vast cometary knots should dissipate. There is, however, just a chance that enough of the dust in some knots will collide and stick together, eventually forming planet-sized bodies. These would be very strange objects indeed: icy bodies like Pluto, but on a massive scale, the size of Earth or even larger. They would not be confined in predictable orbit around the dead star, but would escape – like the rest of the matter in the nebula – into interstellar space, wandering in the dark for ever.

This presents an intriguing and slightly unnerving prospect. We can see the formation of cometary knots in the Helix nebula only because it is (comparatively) close to us; but there is no reason to suppose that this is a unique occurrence, and there may well be cometary knots emerging in all the other planetary nebulae that we simply cannot spot because of their sheer distance. If so, then space could be littered with billions upon billions of massive 'icebergs' flying around in all directions. If humans ever take up interstellar space flight, they would be well advised to keep a sharp lookout. Collision with one of these would be just as fatal as a real Arctic iceberg was for the *Titanic*.

Once a planetary nebula has been blasted out into space, all that remains is a slowly cooling white dwarf which, as we have seen, should eventually fizzle out into a cold, dead, black dwarf. If current theories about the rate of star cooling are correct, the universe is not yet old enough for any white dwarfs to have finally turned black. But observations with Hubble will provide a way of experimentally checking the theory, and, incidentally, providing an independent measure of how old the universe really is.

A team from the University of British Columbia in Canada led by Professor Harvey Richer have been looking at a population of white dwarfs some 7,000 light-years away in a 'globular cluster' in the constellation of Scorpius. These clusters are, in effect, star graveyards, almost spherical groupings of ancient stars left over from the very earliest stages in the formation of the galaxy.

There are only about 150 globular clusters in the Milky Way. When the galaxy first condensed from a cloud of gas, it was roughly spherical. Then, as the collapse proceeded, it began to rotate, forming a flattened disk; but some, locally dense sections remained with a more nearly spherical distribution: the globular clusters. They contain millions of stars of various sizes, but the stars in each cluster are all the same distance from Earth and were all formed at the same time. These two properties make globular clusters immensely important objects for astronomers to study. Professor Richer and his team focused on the nearest one to Earth, known as M4 (i.e. the fourth object in the catalogue compiled by the French astronomer Charles Messier in 1774). M4 is so ancient that all the stars in it with a mass comparable to that of our Sun are expected to have gone through the red giant phase and collapsed to white dwarfs by now. Some estimates are that the stars in M4 are as much as 14 billion years old, whereas estimates for the age of the universe

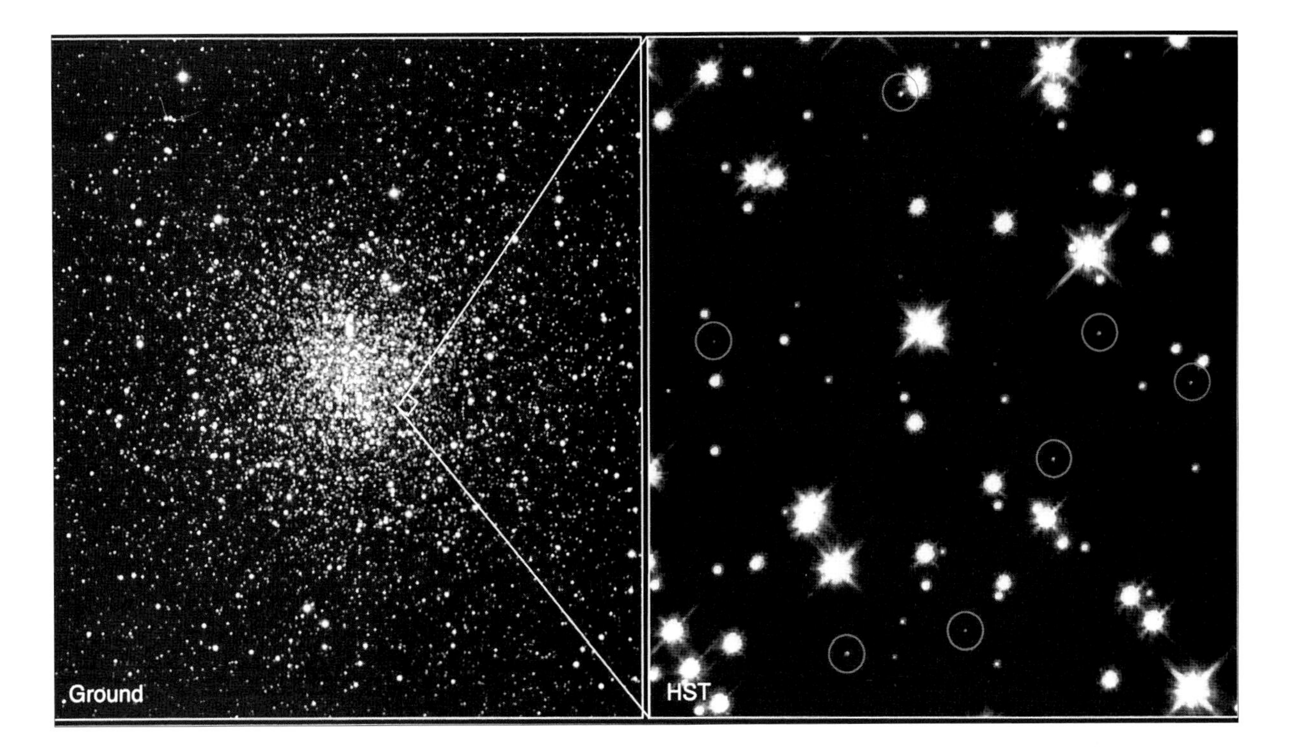

itself have ranged from a mere 8 billion years to 20 billion years. Clearly, if M4 is shown to be 14 billion years old, the universe must be considerably older than that. But how to measure the age of M4? That is where the white dwarfs in the cluster come in.

The Hubble Space Telescope can pick out individual white dwarf stars (circled) from the brighter, sun-like stars and cooler red dwarfs in this close-up (right) of the globular cluster of stars M4. The ground based image (left) represents a field about 47 light years across, whereas the Hubble close-up is a mere 0.63 light years across. M4 is about 7,000 light-years away and contains more than 100,000 stars of which 40,000 are thought to be white dwarfs.

Credit: Kitt Peak National Observatory 0.9-meter telescope, NOAO and M. Bolte (University of California, Santa Cruz). H. Richer (University of British Columbia) and NASA.

The entire cluster is expected to contain about 40,000 white dwarfs. Using a five-hour exposure of the Wide Field and Planetary Camera 2 on board Hubble, Professor Richer and his team found seventy-five white dwarfs in one small region of M4. The light reaching Hubble from the brightest white dwarf is equivalent to the light given out by a 100-watt light bulb placed on the Moon 239,000 miles (385,000 kilometres) away. The faintest was one-fortieth as bright. By carefully examining the light from these stars, the astronomers can deduce the temperature of the surfaces of the stars and hence get an idea of how fast they are cooling, and how long the oldest of them have been in white dwarf mode. The Hubble Space Telescope is thus keeping astronomers busy calculating the answer to one of the most fundamental questions of them all: how long has the universe been around?

Not all stars are destined to end up as white dwarfs, slowly fading away. Other stars definitely go out with a bang and not a whimper. Stars about ten times as massive as our Sun burn their fuel 5,000 times more quickly, and they run through their hydrogen fuel in just 20 million years or so (compared to the 10 billion years for the Sun). These stars live fast and die young. They burn out more quickly precisely because they are so massive: the

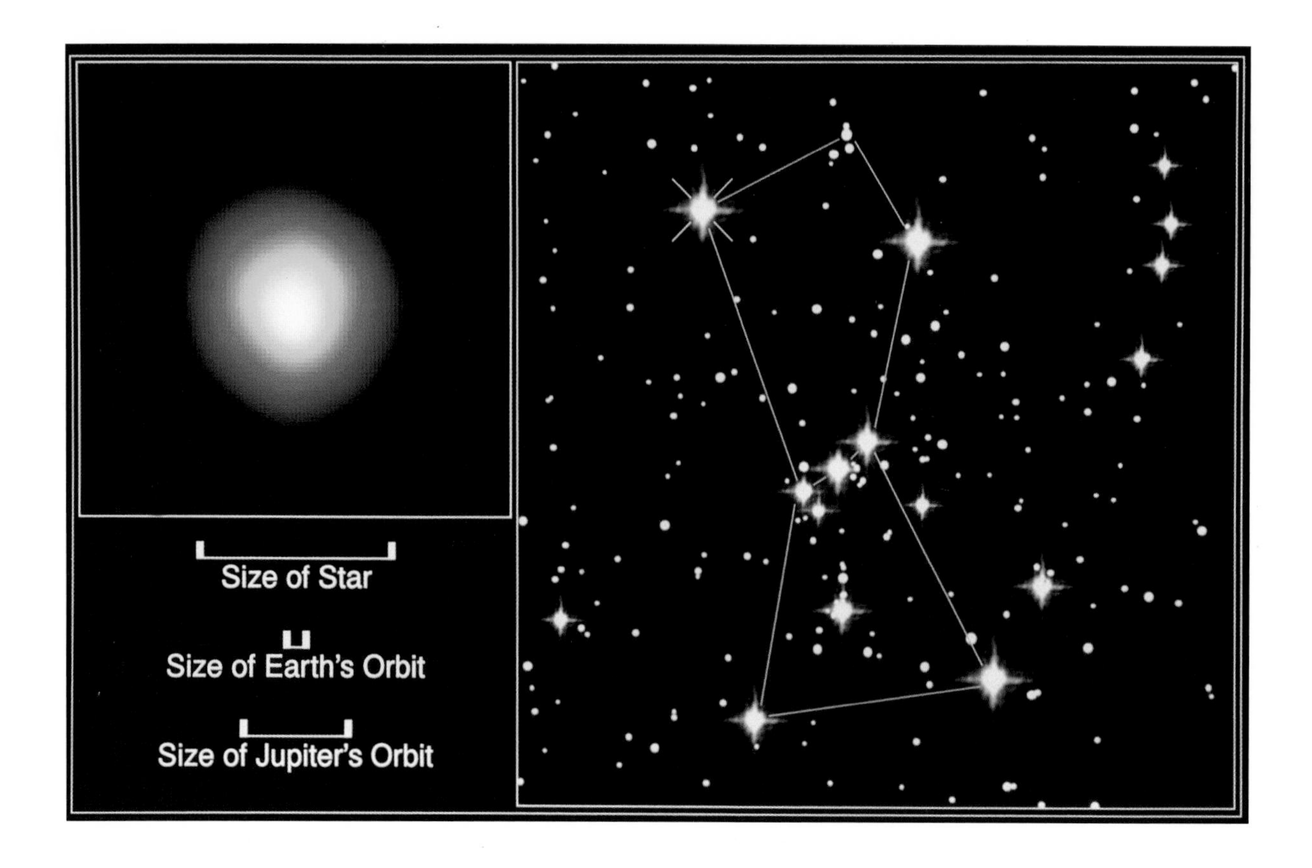

A direct image of Betelgeuse (Alpha Orionis), a variable red supergiant in the constellation of Orion the Hunter. It is about 650 light years away and its apparent size, from Earth, is 20,000 times smaller than the full Moon. To image its disk in this way is like being able to distinguish the two headlights of a car 6,000 miles away. The star has a huge atmosphere glowing with ultraviolet light and a bright spot ten times the diameter of Earth, at least 2,000 Kelvin degrees hotter than the surface of the star. (This is a false colour representation of an image taken in ultraviolet light.)

Credit: A. Dupree (Harvard-Smithsonian CfA), R. Gilliland (STScI), NASA and ESA

core is squeezed more tightly and so the hydrogen fuses into helium more efficiently. Because they are so hot they start out blue, but they too change to a red phase as the hydrogen runs out in their core, except that they become not red giants but red supergiants. The helium fusion into carbon takes over as with red giants, but this time, instead of choking the core, the carbon itself is transmuted into further elements – magnesium or neon. The intense temperatures and pressures of the interior of the star promote further fusions: carbon with helium to form oxygen, which in turn can be fused to form silicon and sulphur. By the end of its life the star consists of a series of layers with different nuclear reactions going on in each: the outer shell is burning hydrogen to make helium; inside that, some recently formed helium is burning to make carbon, oxygen and neon. Further inside are layers of magnesium, silicon and sulphur.

The process continues until iron and nickel build up in the core of the star. Unlike all the other, lighter elements, when iron fuses it takes in rather than gives out energy, so eventually the interior powerhouse of the star is shut off. Without power, the core collapses to the density of an atomic nucleus under the force of its own gravitation. The process takes about a second. This sudden collapse

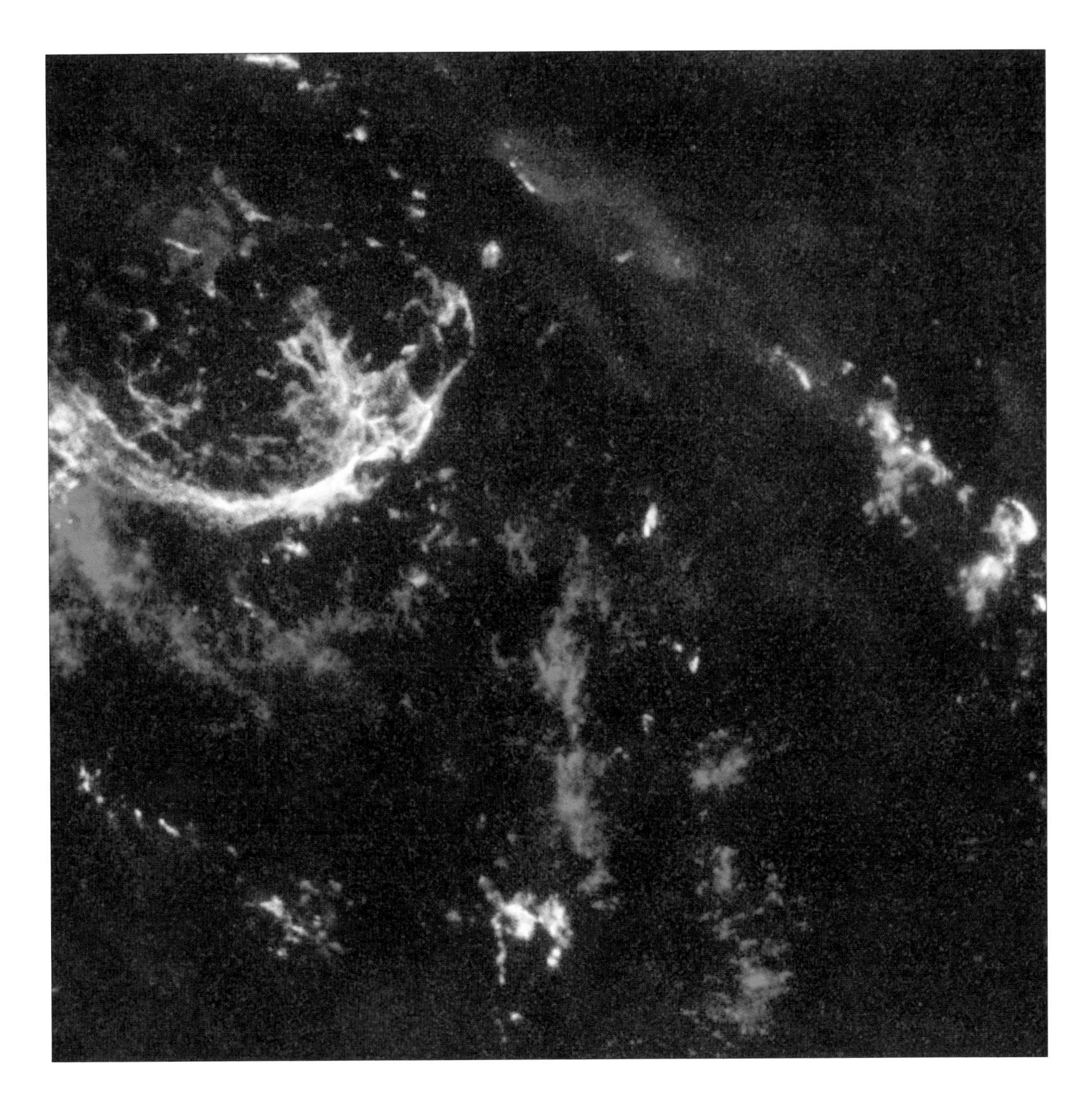

releases an immense blast of energy, equal to about ten times the energy the star produced during its lifetime by burning hydrogen to helium. Gravity and not nuclear fusion, it turns out, is the ultimate powerhouse for massive stars, responsible for more than 90 per cent of their energy output – but all in the final second of the star's existence.

A supernova explosion is one of the most spectacular events in the universe. Most astronomical

This supernova remnant, N132D, represents the still-glowing debris blasted into space when a star blew itself apart 3,000 years ago. The supernova explosion happened 169,000 light-years away in the Large Magellanic Cloud, a satellite galaxy to our own. The blue-green filaments correspond to oxygen-rich gas ejected from the core of the star. Dense interstellar clouds that surrounded the exploded star glow red as the shock wave from the supernova compresses and heats them up. The explosion would have been seen as a new star, 1,000 BC.

Credit: J. Morse (STScI) and NASA

processes involve immense energies – often beyond ordinary human comprehension – but they tend to proceed at a pace that makes the average snail look like an Olympic sprinter. But a supernova happens on a very human timescale – astronomers can watch it as it happens. The blast from the collapse of the core releases so much energy that the star expands rapidly; at the rate of 6,000 miles (10,000 kilometres) a second the star can grow to the size of the entire solar system in just one day. The surface area, and therefore the brightness, of the star will increase many millions of times as it expands. For a week or more a supernova may outshine all the other stars in its galaxy and be some 10 billion times brighter than the Sun.

At this point, some details become hazy, even for astronomers. They have employed some of the world's largest supercomputers to try to calculate the last moments of the supernova and, in particular, to work out how the collapse of the core can cause the other layers of the star to expand. They have not yet been completely successful. (One recent supernova, 1987A, went on from a red to a blue supergiant phase before exploding, although that has been reconciled through a slight revision of theory.) The complexity is not diminished by the fact that most of the energy is released in the form of radiation, consisting of exotic subnuclear particles called neutrinos. Astronomers think that the core collapses so quickly that it overshoots and bounces back, collides with the infalling gas and creates a shock wave that sends the remainder of the star flying outwards, possibly with the assistance of radiation pressure from the neutrinos.

The general picture, however, is clear and accurate and has been confirmed by many observations. The outer layers are blasted off into space in an explosion that releases as much energy as was generated during the entire previous lifetime of the star (this of course does not count the other 90 per cent of the energy released by the collapse which has been carried off by the neutrinos). In those last few moments, the temperature rises in the regions beyond the core so that fusion does proceed, and elements heavier than iron are generated in the exploding layers of the supernova and dispersed to interstellar space – everything from gold to uranium.

It is worth pausing for a moment to consider how important this process is for life. The red colour of our blood stems from haemoglobin, the molecule that carries oxygen to our muscles and removes the spent carbon dioxide. Central to the haemoglobin molecule is an atom of iron; if illness or dietary deficiency reduce our iron intake then we suffer from anaemia. The iron in our blood's haemoglobin, so essential for our health, was forged in the dying moments of a massive star that went supernova many billions of years ago. As it flared up in a brief moment of radiant glory and then exploded, it shot out all the new chemical elements it had created into interstellar space. Those elements later came together as part of the protoplanetary disk that spawned the Earth. The very blood in our veins is the dust of stars long dead.

Two supernovae contributed to the end of the old system of Ptolemaic astronomy. In 1572 (before the invention of the telescope) the Danish astronomer Tycho Brahe discovered a supernova in the constellation of Cassiopeia and showed that it had to be more distant even than Saturn. Since Ptolemaic theory had it that there could be no changes among the fixed stars – for they were made of eternal stuff and not the temporal, changeable material here on Earth – Tycho Brahe's observations provided an irreconcilable contradiction to the old theory. This was compounded when, thirty-two years later in 1604, Johannes Kepler (Brahe's student) also discovered a supernova. The stage was set for the later and final confrontation between Galileo and the Church, in which the Church was apparently the victor by confining Galileo to house arrest, but which in the long run it lost comprehensively.

Supernovae are relatively rare. Whereas astronomers expect to see an ordinary star like our Sun turn into a red giant about once a year in the Milky Way, supernovae are expected to occur only once every fifty years or so in a galaxy like ours. Occasionally, supernovae are so bright that they can be seen with the naked eye on Earth; there have been five such visible supernovae in the past millennium. The most recent was in 1987.

Astronomers have over decades built up complex models of how they expected the shock waves caused by supernova explosions to behave. Hubble's fine resolution allowed them, for the first time, actually to see the shock waves in action, and thus to compare the reality with the theory. One image, of part of the Cygnus Loop, the remnants of a supernova explosion that occurred about 15,000 years ago, shows the effects of the blast crashing into tenuous clouds of interstellar gas. The force of the collision compresses and heats the gas, making it glow – and allowing us to see it.

The Hubble Space Telescope has been observing not only the aftermath of supernova explosions but some massive stars that appear to be in the final stages preparatory to going supernova. One such star, Eta Carinae, has been known for many years from ground-based observations. This appeared to be just an ordinary star in the southern hemisphere until 1833 when it began to vary erratically; at its brightest, only Sirius the Dog Star outshone it, despite the fact that Eta Carinae is about 8,000 light-years distant (Sirius by contrast is a next-door neighbour, just 8.7 light-years away). Eta Carinae is about 100 times as massive as our Sun, and before it lapsed again was up to 4 million times brighter. Its great flare-up has already blown off some material to form a nebulous patch around the star, known as the Homunculus nebula. Its likely fate, perhaps only in a few hundred years from now, is to become a supernova. It is a star under sentence of death.

Although Eta Carinae can be observed from the ground, the Hubble Space Telescope has provided unprecedented details of the spectacular double cloud of gas and dust billowing out from the star. As with the red giants, the puzzle of a lack of symmetry is present here too, with two polar lobes and a central disk of dust around the star's equator. The material is moving outwards at a speed of more than 100,000 miles (160,000 kilometres) an hour. At the centre there is intense emission of ultraviolet light, whereas the lobes contain large amounts of dust which absorb the ultraviolet and appear to be more reddish in colour. The outer regions are around 100,000 times fainter than the centre, and this image is the superposition of eight exposures in a sequence of red and near-ultraviolet light.

And when astronomers used the Goddard High-Resolution Spectrograph carried aboard the telescope, they got a real surprise. They discovered that the clouds surrounding Eta Carinae are acting as an ultraviolet laser, producing rays shooting out of a dying star. The laser light was stubbornly invisible to ground-based telescopes because the Earth's atmosphere absorbs all radiation at these wavelengths. Only from its vantage point above the atmosphere was the Hubble Space Telescope able to see the light. Neon and ruby lasers are now commonplace here on Earth, not only as laboratory instruments but as ordinary domestic items: compact disk players use a small solid-state laser to 'read' the pattern of information encoded on the surface of the CD. However, an international team of Swedish and US astronomers have found that the source of the ultraviolet laser is highly energetic ionised atoms of iron in the clouds surrounding Eta Carinae. Nothing like it has been seen before anywhere.

Until October 1997 Eta Carinae had been thought to be the most massive and intrinsically

This may well be the most luminous star in our galaxy – yet, paradoxically, it has remained hidden behind obscuring layers of dust until Hubble's infra-red vision laid it bare. The Pistol nebula that surrounds it, which appears magenta in the picture, is formed of two enormous shells of material blown off the dying star 4,000 and 6,000 years ago. Despite its great loss of material, the star is still shining with the radiance of 10 million Suns.

Credit: Don F. Figer (UCLA), and NASA

brightest star in the Milky Way. Then astronomers announced that an image taken by Hubble had revealed an even more brilliant star at the core of the Milky Way. This star is 25,000 light-years away, in the direction of the constellation Sagittarius towards the centre of the galaxy. It is hidden behind thick clouds of dust and so could only be seen in infra-red light, even though it is shining with the power of 10 million Suns. Astronomers estimate

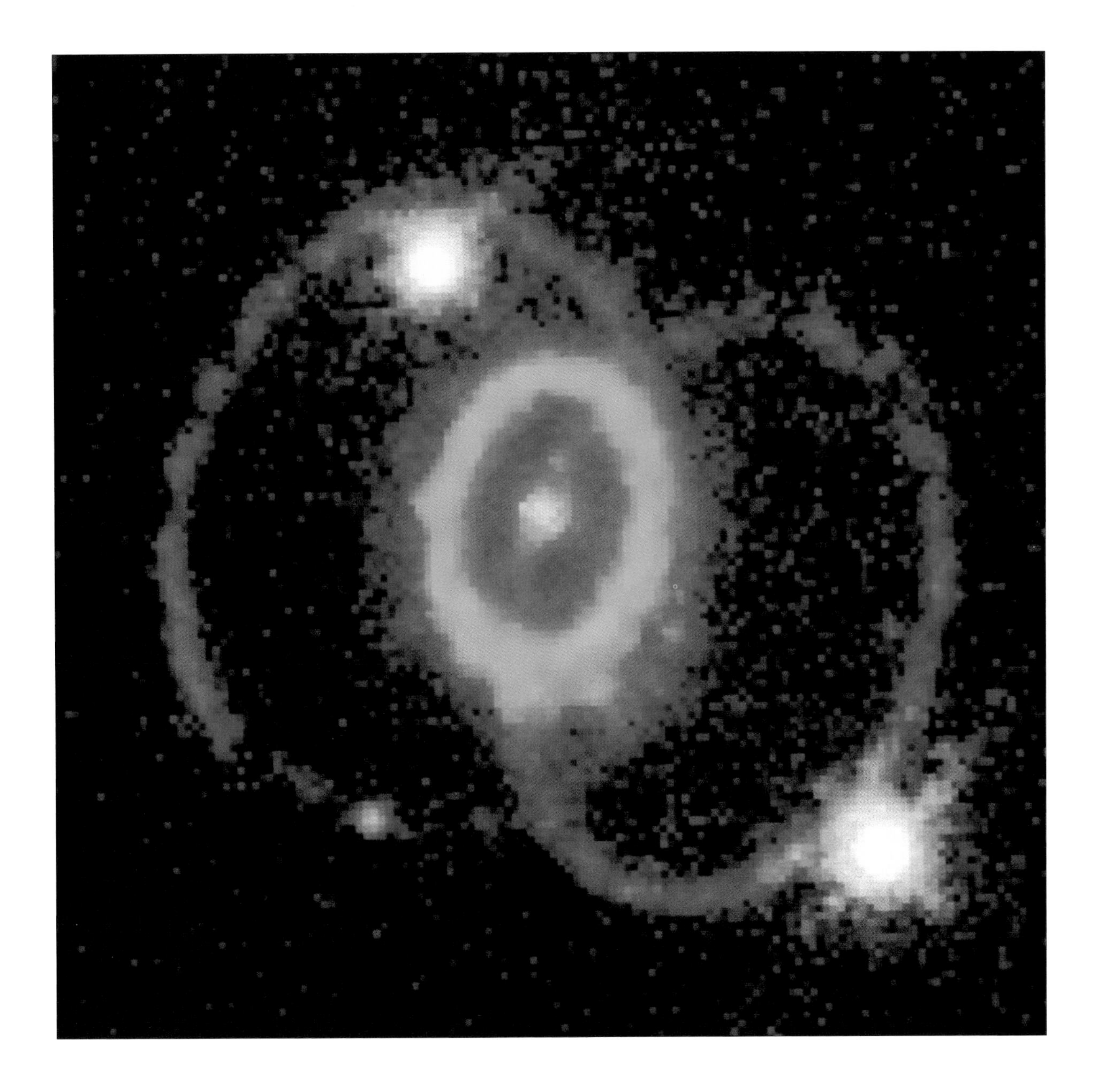

that originally it may have been 200 times as massive as the Sun, but it has been shedding vast quantities of gas and dust and so it may now be 'only' 100 times as massive. It has blown off at least two shells of gas and dust, the largest of which is nearly four light years in diameter, so it would stretch from the Sun to the next nearest star. These two events are recent, by astronomical standards: the shells of gas are just 4,000 and 6,000 years old.

A nest of haloes, formed by rings of glowing gas, surrounds the site of the supernova explosion 1987A – the first supernova to be visible to the naked eye for nearly 400 years. The rings are probably on the surface of a bubble formed by a cloud of slow-moving gas ejected when the star was a red supergiant. A much faster 'wind' of gas followed during the subsequent blue supergiant stage and this escaped at the poles of the star, thus carving the bubble into an hour-glass shape.

Credit: C. Burrows (ESA and STScI), WFPC 2 Investigation Definition Team, and NASA.

Although Eta Carinae and the bright star towards our galaxy's centre are puffing gas and dust away into space, neither has yet flared up into the final gargantuan explosion of a supernova. In February 1987 the first supernova to be visible to the naked eye since Kepler's supernova of 1604 exploded violently. Unbelievably, for an object visible to the unaided eye, Supernova 1987a was not even in our own galaxy, but in the Large Magellanic Cloud, the nearest galaxy to our own, and which is some 160,000 light-years away.

In 1994 Hubble obtained spectacular images of three concentric rings surrounding the supernova. The angle from which the image was taken is slightly deceptive and although the rings look as if they intersect each other they are probably in three different planes. The smaller, bright ring is probably in the same plane as the supernova itself, with the other two in front of and behind the exploding star. But how have those two rings been illuminated enough that they can be seen?

Astronomers speculate that they are traced out by a high-energy beam of radiation making a circle in the 'sky' above the supernova. Where the beam touches gas and dust ejected by a star, it will heat them until they glow brightly. To create searchlight beams to trace out the outer rings, astronomers think that there must be not only the remnant of the supernova itself but a companion star. The gravity of the supernova remnant would suck material down on to itself from the companion, which would then be heated and blasted back into space along narrow jets, along with extremely thin beams of radiation. Dr Chris Burrows of the European Space Agency, who led the team that took the picture, described Supernova 1987a as 'an unprecedented and bizarre object. We have never seen anything behave like this before.'

A couple of years later, Hubble conducted a chemical analysis of the material in the inner ring, and picked up tell-tale traces of oxygen, nitrogen and sulphur. It was also able to detect the consequences of the collision between the hot, fast-moving gas ejected in the supernova explosion and the cooler, denser gas blown off the outer parts of the star some 20,000 years or so earlier. The spectrograph picked up signals from hydrogen travelling at 33 million miles per hour (15,000 kilometres a second). Travelling at this speed, the gas would cross our solar system in just a day. In addition to hydrogen, signals from nitrogen and helium gas were also detected.

One of the most beautiful objects in the sky, the Crab nebula, is the result of a supernova explosion, one which was visible in 1054; indeed, it was so bright that it could be seen during the daytime. Chinese astronomers recorded the event, and concluded that it was a 'guest' star in what we now describe as the constellation of Taurus the Bull. The cloud of gas that originated in that stellar explosion almost 950 years ago has now expanded more than a million-fold in diameter. The cloud can be expected to shine for tens of thousands of years, but studies made using Hubble have shown that the cloud is changing subtly over periods as short as a few days. A 'wind' of subatomic particles is still being accelerated out from the supernova remnant at the centre of the nebula, and when this slams into the slower-moving envelope of gas and dust blown off in the original explosion, then the wisps of gas dance and wriggle in response.

The remnant left behind from a supernova explosion is among the most exotic of all the deeply exotic inhabitants of our universe. The original star is so extraordinarily massive that when its core collapses, triggering the explosion of the outer layers, the core itself contracts below the limit of a white dwarf and becomes instead a neutron star, the densest and tiniest stars in the universe (with the possible exception of a black hole, of which more later).

Neutron stars have gone beyond the limits of normal materials and are rather like a single giant atomic nucleus, except that they can be about 12 miles (20 kilometres) in diameter and are encased in iron armour. As the stellar core contracts, the electrons which normally orbit the protons and neutrons in the centres of the atoms are forced into the atomic nuclei themselves. There they combine with the protons to form more of the neutral sub-atomic particles, neutrons. The density of a neutron star is about the same as that of an atomic nucleus, i.e. about 100 million million times as dense as water. One teaspoonful of a neutron star would weigh about 100 million tonnes. The outer surface is not an ionised gas, like other stars, but a sheet of iron. Under the impact of the star's gravity and its intense magnetic field, the iron polymerises to form a material 10,000 times as dense as Earthly iron and with a strength 1 million times that of steel. Inside, as the pressure increases, ordinary matter is squeezed out of shape and a nuclear 'sea' of neutrons forms. It is a superfluid: it flows and moves without experiencing any resistance or viscosity.

Although neutron stars are truly exotic objects, many of their properties were worked out in the late 1930s, more than sixty years before the first concrete evidence of their existence. One of the leading scientists who developed the theory of neutron stars was J. Robert Oppenheimer, the man who later, during the Second World War, became the 'father' of the American atomic bomb.

Neutron stars spin on their axis with intense rapidity. The one in the Crab nebula rotates thirty times a second. As they spin they emit great search-light-like beams of radiation, – rather like the beams from a stellar lighthouse. Quite how they do this is, as yet, unexplained. Most pulsars emit energy mainly in the radio wavelengths, rather than in visible light; only the pulsar at the heart of the Crab nebula and one at the core of the Vela nebula are known to emit pulses of light. In October 1996 Hubble found a lone neutron star somewhere between the Earth and the southern constellation Corona Australis 400 light-years away. It was the first direct look in visible light at a neutron star.

There is, of course, one stage further in the evolution of some stars. Those that are more than thirty times as massive as our own Sun – like Eta Carinae – will follow the supernova path. But as their cores contract, their sheer mass means that gravity pulls them in beyond the stage of white dwarf and then neutron star. They carry on collapsing until they disappear altogether. This state of ultimate annihilation is what is known as a 'black hole'. The gravity is so intense that not even light, the fastest thing in the universe, can reach escape velocity to get out into space from the surface of such a black hole. Detecting such stars – which are not so much dead as non-existent – might be impossible, were it not for the fact that their intense gravity sucks in material from space, and as this dust and gas spirals in down on the black hole, it will emit radiation, usually in the form of pulses of X-rays. The role of the Hubble Space Telescope in detecting such objects – predicted in Einstein's General Theory of Relativity – will be described in the next chapter.

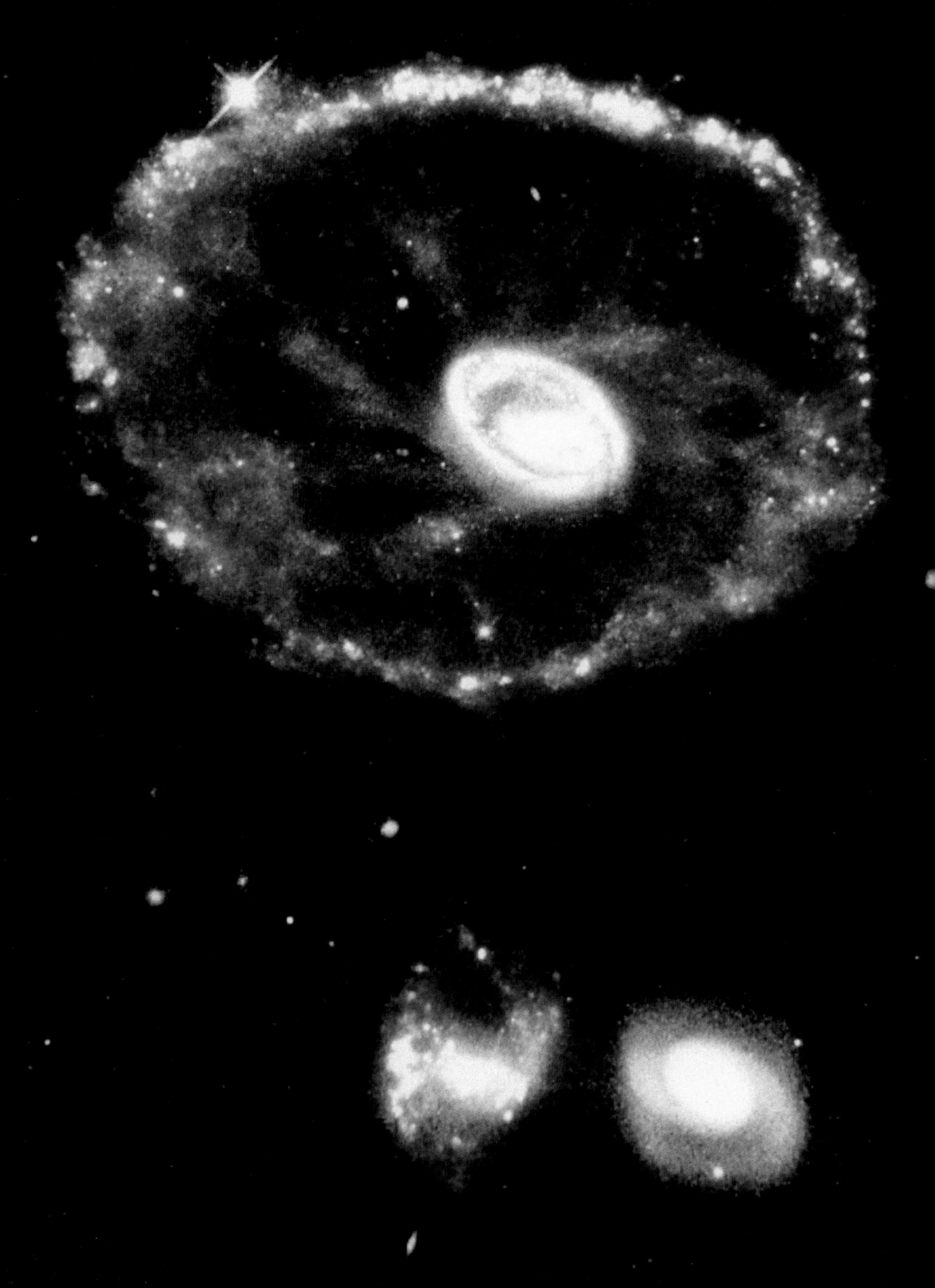

CHAPTER FIVE

THE CITIES OF THE STARS

Astronomy is among the oldest of the professions. About 3000 BC, the ancient peoples of Babylon, Egypt, India and China started observing the stars and planets in order to construct calendars for the regular planting of crops. They were carrying out sophisticated astronomical observations while the nomad inhabitants of Europe were barely scratching a living herding their sheep and goats from place to place. Civilisations came and went, but all the while men and women continued to observe the stars.

Opposite: The Cartwheel galaxy is the result of a head-on collision between two galaxies. The intruder may be one of the two galaxies shown nearby. It is unlikely that any stars actually bumped into each other, but like a stone thrown in a pond, the gravity of the intruder sent density waves rippling through the gas and dust in front of it, producing the outer blue ring around the galaxy glowing with billions of new stars. The Milky Way could fit within the ring, which is 150,000 light years across. The galaxy's spiral structure is beginning to re-emerge, as in the faint arms extending out from the core.

Credit: K. Borne (ST ScI), and NASA

Above: A close-up of the core of the Cartwheel galaxy shows vast quantities of dust with bright pinpoints of light from stars forming in giant clusters, again kicked into life by the collision that probably occurred 200 million years before. Galaxies travel through space at speeds approaching 2 million miles per hour and so collisions do occur. The galaxy is in the constellation Sculptor, visible in the southern hemisphere, and it lies some 500 million light years from Earth. This image and the one opposite are a combination of exposures: an hour-long exposure in blue light and half an hour in near-infra-red light

Credit: K. Borne (ST ScI), and NASA

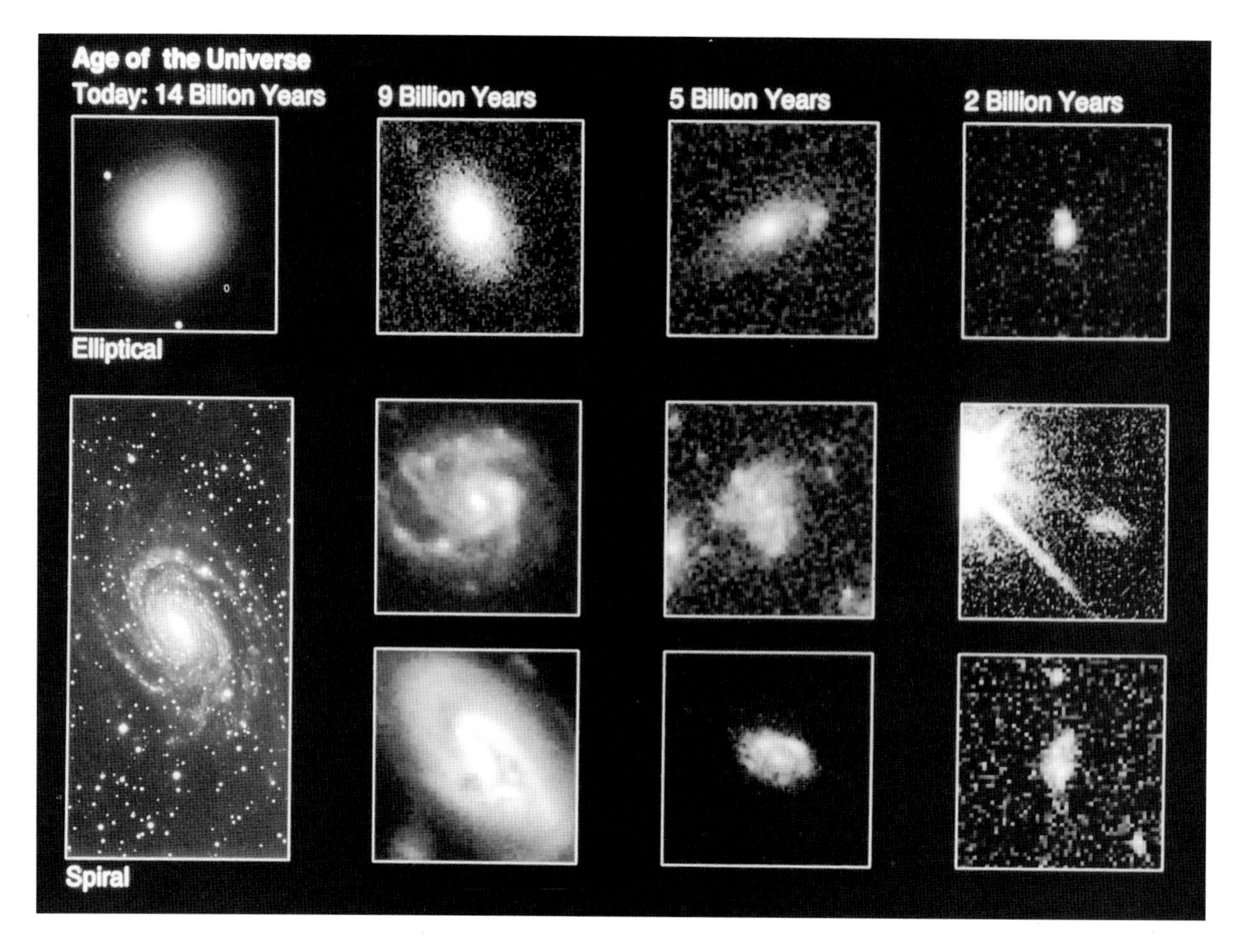

Far left column: Nearby galaxies show typical spiral and elliptical shapes characteristic of the current stage in the universe's evolution. Elliptical galaxies contain older stars, while starbirth is still going on in the outer arms of the spirals.
Centre left column: When the universe was about two-thirds its present age, elliptical galaxies appear fully evolved and resemble present-day models while some spirals are more loosely shaped than the present day version.
Centre right column: The more distant the galaxy, the further back in time is the image that we get of it. Spiral structures appear even vaguer for galaxies representative of how the universe was at one-fifth its present age. Ellipiticals (*top*) are still just recognisable as such, however.
The far right column shows extremely remote objects as they were when the universe was around one-tenth its present age. The object in the top frame has the light profile of a mature elliptical, implying that they must have formed remarkably early whereas spirals were slower to develop.

Credit: A. Dressler (Carnegie Institutions of Washington), M. Dickinson (STScI), D. Macchetto (ESA/STScI), M. Giavalisco (STScI), and NASA

Occasionally, the light of knowledge went out as this or that barbarian horde overran some cultured city. Sometimes it was not even barbarians: the Romans were responsible for an irreparable loss when they allowed the great library of Alexandria to burn to the ground. Sometimes it was the otherwise civilising influence of the Christian religion that snuffed out the light; its adoption of the inflexible Ptolemaic system of astronomy ended up in the loss of the wisdom of the classical Greeks.

Some knowledge was lost but regained many centuries later. One characteristic of the Dark Ages that followed the collapse of the Roman Empire was a shrinking of perspective. Once it had been possible to travel from England to Egypt under the protection of the Pax Romana. For centuries thereafter, Europe split into small kingdoms each regularly at war with its neighbours and many at war within themselves. Horizons shrank to the kingdom, the region, the county, the village even.

The horizons of scholars shrank too. In about the year 200 BC, the Greek mathematician Eratosthenes worked out not only that the Earth was round, but calculated its circumference accurate to within 10 per cent of its modern value of 24,900 miles (40,074 kilometres). He did so by comparing the length of the shadows cast on midsummer's day at two Egyptian towns whose distance from each other he had previously measured. But after the passing of the Roman Empire, scholars pictured the Earth as a flat surface not much bigger than the lands surrounding the Mediterranean Sea. Even when, with the Renaissance, men began again to think of the Earth as round, they could not image its true scale. Perhaps the most famous miscalculation in history is that of Christopher Columbus. He reckoned that the coast of Asia would be just 3,900 miles (approximately 6,300 kilometres) west of Spain. It was a fortuitous mistake, for if he had known the real distance Columbus presumably would never have set sail, and so would not have reached America.

Galileo, building on the work of his predecessors Kepler and Tycho Brahe, set in motion the revolution in humanity's understanding of the scale of the universe in which we live. But a further 300 years were to elapse before the era of truly modern astronomy began, and with it an understanding of how we, just like Christopher Columbus, had grievously underestimated the size of the universe.

The scale of our own solar system seems vast enough. If it had been possible for a latter-day Columbus to board the unmanned Pioneer or Voyager probes on their journeys to the outer limits of our solar system, his trip would have lasted more than a decade – and that was with the benefit of hitching a free 'lift' from the gravitational pull of some of the planets, which acted in slingshot fashion to speed the probes on their way. A return journey then, were it possible, would occupy something more than twenty years. Human technology is in fact nowhere near capable of such a feat for the foreseeable future. No human has set foot on the Moon for more than a quarter of a century, and the idea of a manned landing on Mars remains a technological fantasy, as planners struggle to cope with the implications for the mental and physical well-being of a crew cooped up together for six months without a break. None the less, the ultimate prospect of humans journeying to (and, one hopes, returning from) the neighbouring planets does not seem impossible, with improvements to current technology.

But beyond Pluto the distances exceed the human scale. As we saw in Chapter Three, to reach the nearest neighbouring star would take a modern spaceship more than 100,000 years. Crossing to the far side of our galaxy would take 2 billion years. And to reach the next large galaxy to our own – Andromeda – would take 30 billion years at the speeds our technology allows today, more than twice as long as the entire universe has already been in existence.

The universe is a lonely place for creatures such as ourselves. Our sense of scale is foreshortened in cosmic terms: because all our subjective experiences are conditioned by life on the surface of the Earth, a small and crowded planet, we have no true feeling for the distances between the galaxies; and because our sense of time is conditioned by our short and eventful lives, it is very hard for us to grasp how long the universe has existed, and how slowly some interactions within it might proceed.

It seems almost an abuse of language to call an object as unfathomably remote as Andromeda our 'neighbour', but on a cosmic scale that is what it is. Andromeda has played a crucial role in modern astronomy. It was the turning point in our understanding of the universe, and the new era of truly modern astronomy can be dated quite clearly to the 1920s, that incredible decade after the publication of Einstein's Theory of Relativity which saw also the invention of quantum mechanics. And the architect of the new astronomy was Edwin P. Hubble.

By the 1920s a great debate was raging in astronomy. Many nebulae could be seen in the night sky. What was their nature? Some astronomers thought they were clouds of dust and gas within the Milky Way, collapsing to form new stars. One of the most prestigious proponents of the view that the nebulae were associated with our own galaxy was an almost exact contemporary of Hubble's and someone who, like Hubble, was working at the Mount Wilson Observatory. Harlow Shapley was the first astronomer to measure the size and structure of the Milky Way, so his views carried weight in the great debate. But others followed the German philosopher Immanuel Kant in thinking that some of the nebulae might be complete star systems outside our own. No one knew for sure.

The confusion had been around for a long time. The French astronomer Messier, working in the mid-eighteenth century, had compiled the first ever catalogue of nebulae because he was searching for comets and wanted to list the fuzzy objects that clearly weren't comets but might be confused with them. The listing was done on the basis of appearance, rather than the true nature of the objects, which was unknown to Messier and his successors until this century. The Messier catalogue thus jumbles quite different objects together. Andromeda, an entirely different galaxy to our own, is object 31 in the Messier catalogue (and hence, designated as M31); and the Ring nebula, which is a planetary nebula within our galaxy, is also listed in the catalogue (as M97).

Nowadays, thanks to better optical telescopes as well as ones that can 'see' infra-red light and radio waves, not to mention space-borne instruments, astronomers know that some nebulae (such as the Orion nebula, which is also known as M42) are indeed stellar birthplaces within our galaxy. They also know that some nebulae are the tombstones of dust and gas remaining from the outer layers of red giants – the inaccurately named planetary nebulae – also within our galaxy. Some, rich in all the chemical elements, are the clouds blasted into space by the explosions of supernovae; rarer events, but still possible within our galaxy. Others still are tightly packed 'globular clusters' of stars; but again, local highly concentrated patches of ancient stars within our own galaxy.

It was only in the 1920s that Edwin Hubble was able to bring the smudge in the sky that was Andromeda into sharper focus than any astronomer had achieved before him. The outer parts of this nebula became visible as stars in their own right.

But even this did not settle the controversy completely; Andromeda could simply have been yet another example of a Milky Way globular cluster. Hubble had to find a way of measuring the distance to Andromeda.

Hubble built on earlier work by the brilliant American astronomer Henrietta Leavitt. She had spent years studying variable bluish stars, known as Cepheid variables. In 1912 she announced that the frequency with which these stars brighten and dim is related to their intrinsic luminosity, so by measuring their rates of variation she could work out how intrinsically bright they were. By comparing their intrinsic brightness with the degree to which they appear dimmer viewed from Earth, she obtained a direct measure of how far distant they are. Hubble applied Leavitt's ideas to the Andromeda nebula, where he found a set of bright blue variable stars very similar to those that Leavitt had detected in and close to our own galaxy.

Astronomers looking at this magnificent Hubble image of galaxies both close and very distant found eighteen small and very young galaxies spread across a relatively small distance. Some speculate that we are looking at the building-blocks of today's larger galaxies, formed partly as a result of collisions and mergers between these closely packed early galaxies.

Credit: Rogier Windhorst and Sam Pascarelle (Arizona State University), and NASA

And it was as a result of using this new way of measuring a star's distance from Earth that Hubble was able to announce in 1924 that the variable stars shining from Andromeda could not possibly be inside the Milky Way. They are 2.2 million light-years away, whereas the entire Milky Way is 'only' 100,000 light years across. (One of the ironies of history is that Hubble actually got the distance

The globular cluster 47 Tucanae lies 15,000 light-years away within the Milky Way in the constellation Tucana. Globular clusters tend to contain older reddish stars but this one has 'blue stragglers' which glow with the blue light of young stars. A ground-based telescope image (inset) shows the entire crowded core. The circles in the Hubble telescope image highlight several of the cluster's blue stragglers.

Credit: R.Saffer (Villanova University), D. Zurek (STScI) and NASA

wrong in 1924. He did not know, and could not have known then, that there are actually two types of Cepheid variables. When this was realised in 1952 – the year before Hubble's death – estimates of the size of the universe were doubled at a stroke.)

But even though he underestimated the true distance, Hubble came to understand that Andromeda is a separate star city, an island universe

They look like slightly fuzzy stars, but in fact they are entire cities – connurbations populated with billions of individual stars – which are so far distant that only the collective light emission can be seen from Earth. The realisation that these were distant galaxies and not nearby fuzzy clouds revolutionised astronomy in the late 1920s. The differing shapes of these ancient galaxies gives clues to their history and their development.

Credit: NASA

in its own right. Very quickly thereafter astronomers realised that most of the fuzzy patches of light – nebulae – that they saw when they looked away from the direction of the Milky Way were extragalactic. The idea that the entire universe consists of many galaxies, so distant that their individual stars cannot be resolved even with the most powerful of telescopes, was a new and profoundly unsettling one.

In 1925 Hubble started to classify the galaxies that he had observed on the basis of their shape. He realised that they fell into four basic categories: elliptical; spiral; lens-shaped (lenticular, as they are known in the trade); and irregular. Over the years since, astronomers have devised many different classifications and subgroups, but Hubble's is still a good guide to the galaxies.

Elliptical galaxies appear through optical telescopes to have smooth ellipsoidal shapes. They appear to have predominantly red stars, old and cool, with little interstellar dust and gas. These galaxies appear to have little new star-formation activity.

In contrast, *spiral galaxies* tend to be slightly smaller but to be more luminous and to have more hot, young blue stars, along with clouds of interstellar dust and gas. Active star formation is still proceeding in these galaxies, mainly in the spiral arms where there is plentiful dust and gas. Our own Milky Way is a spiral galaxy, as is Andromeda. (Ironically, the fact that Andromeda is a spiral shape had been observed some forty years before Hubble proved that it was in fact a galaxy. Many other 'spiral nebulae' had been catalogued in the four decades or so before their true nature was known.)

Lenticular galaxies are so called because they appear to be lens-shaped when viewed edge-on, and they appear to be an intermediate shape between ellipticals and spirals. They do not have the typical spiral arms, but they are flattened like the Milky Way and Andromeda, rather than being rounded and bulked out like the ellipticals.

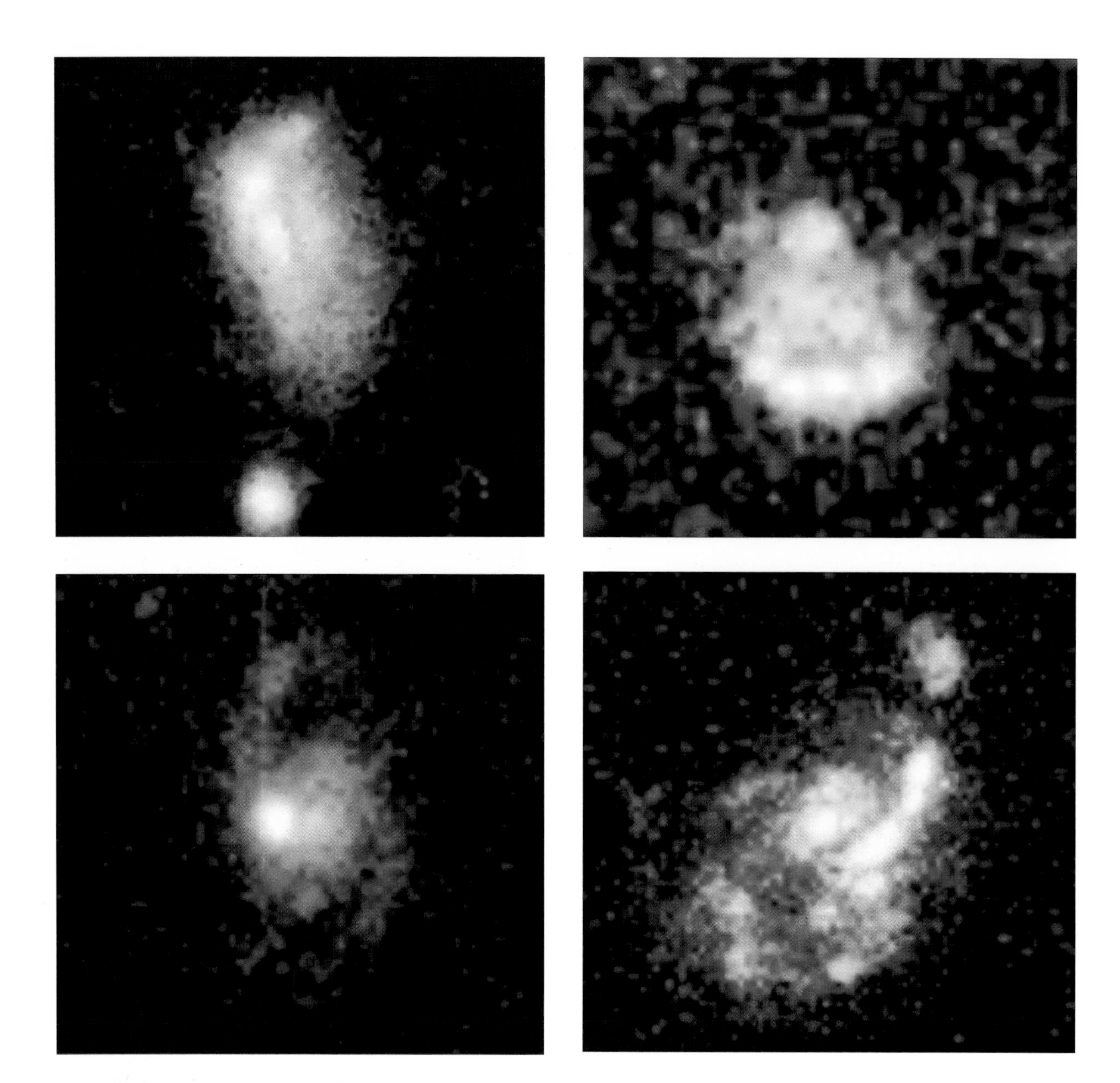

A medium-deep survey by the Hubble Space Telescope of galaxies with peculiar and irregular shapes. These are ancient images of galaxies when the universe was very young. The bright blue areas indicate regions where rapid star formation is taking place. Viewed from the grounds, these distant early galaxies appear as fuzzy, indeterminate blobs. Viewed with Hubble, they are resolved into distinct – and distinctive – shapes. Once, these oddball galaxies were far more common than 'normal' shaped galaxies; but they gradually disappeared, either fading away or being merged with larger galaxies.
Credit: Richard Griffiths (Johns Hopkins University), the Medium Deep Survey Team, and NASA

Irregular galaxies are simply those whose shape is not well enough defined to fit into the other categories. They appear as an amorphous mass of stars, some of them as if they are being disrupted by inner forces or perhaps by the gravitational attraction of a nearby galaxy.

Astronomers know that all the galaxies started to form at about the same time, shortly (in astronomical terms) after the Big Bang that started the whole universe off. Elliptical galaxies appear to have undergone rapid star formation which used up all the available interstellar gas and dust quite quickly. By contrast, spiral galaxies like our own appear to have two phases of starbirth, first in the centre where the composition and age of stars now resemble elliptical galaxies, and then in the dust and gas clouds of the spiral arms, where active starbirth is still going on. The shape of the galaxies does not just offer a means of classification but is related to the dynamics of the galaxies themselves.

The classification of galaxies was but one of three fundamental discoveries made in the decade or so after Hubble's original proof that the Andromeda nebula was an 'external' galaxy.

The second discovery came when it was observed that galaxies were not evenly spread across the sky, but were clustered together. Our own 'local group' consists of about thirty galaxies, dominated by the twin giant spiral galaxies of the Milky Way itself and of M31, the Andromeda galaxy. Our local group is about 7 million light-years across. In contrast, some 50 to 60 million light-years away lies the Virgo cluster, which contains more than 2,000 galaxies. Three of the biggest elliptical galaxies in the Virgo cluster each measure 2 million light-years across – the distance separating the Milky Way from Andromeda.

The third major discovery made about galaxies by Hubble was that they were moving apart at high speed, leading him to conclude that the entire universe was expanding.

Once astronomers had grasped that the universe was fundamentally different from what they had imagined, the next step was to start to make detailed observations about the stars contained within other galaxies.

Within our own local group two notable galaxies are the Large and the Small Magellanic Clouds. These are irregularly shaped, smaller galaxies, and are in fact the nearest galaxies to our own, being only about one-tenth as far away as Andromeda. They are visible to the unaided eye in the southern hemisphere, and were once called 'Cape Clouds' by mariners who observed them as they circumnavigated the Cape of Good Hope at the tip of southern Africa. They were also described by the great explorer Magellan, and it is after him that they are now named. The significance of the Magellanic Clouds is that they are sufficiently close for large telescopes to focus on individual bright stars, while being far enough away for all the stars to be considered roughly the same distance away from us. The Large Magellanic Cloud is about 160,000 light-years distant, whereas the Small Magellanic Cloud is slightly further at about 185,000 light-years. By looking at processes going on within the Magellanic Clouds – and within Andromeda – astronomers can observe events very similar to those taking place within our own Milky Way, but with the added advantage of a sense of perspective. It is sometimes difficult to assess the significance of events if they are close by; better then to witness similar happenings from afar. And the resolution achieved by the Hubble Space Telescope has allowed scientists to make observations of unparalleled clarity.

In 1994 the Hubble Telescope witnessed the effects of a bubble of hot gas from a thousand or more supernova explosions sweeping through part of the Large Magellanic Cloud, triggering the birth of new stars. There is one significant difference between the Clouds and our own Milky Way: they are deficient in the heavier elements compared to the concentrations found in our own vicinity. Elements heavier than helium are created in the interiors of stars, and then spread across interstellar space by the explosions of supernovae, eventually to be recycled in a new generation of stars. The comparative deficiency of heavier elements in the Magellanic Clouds suggests that the rate at which matter has been processed through

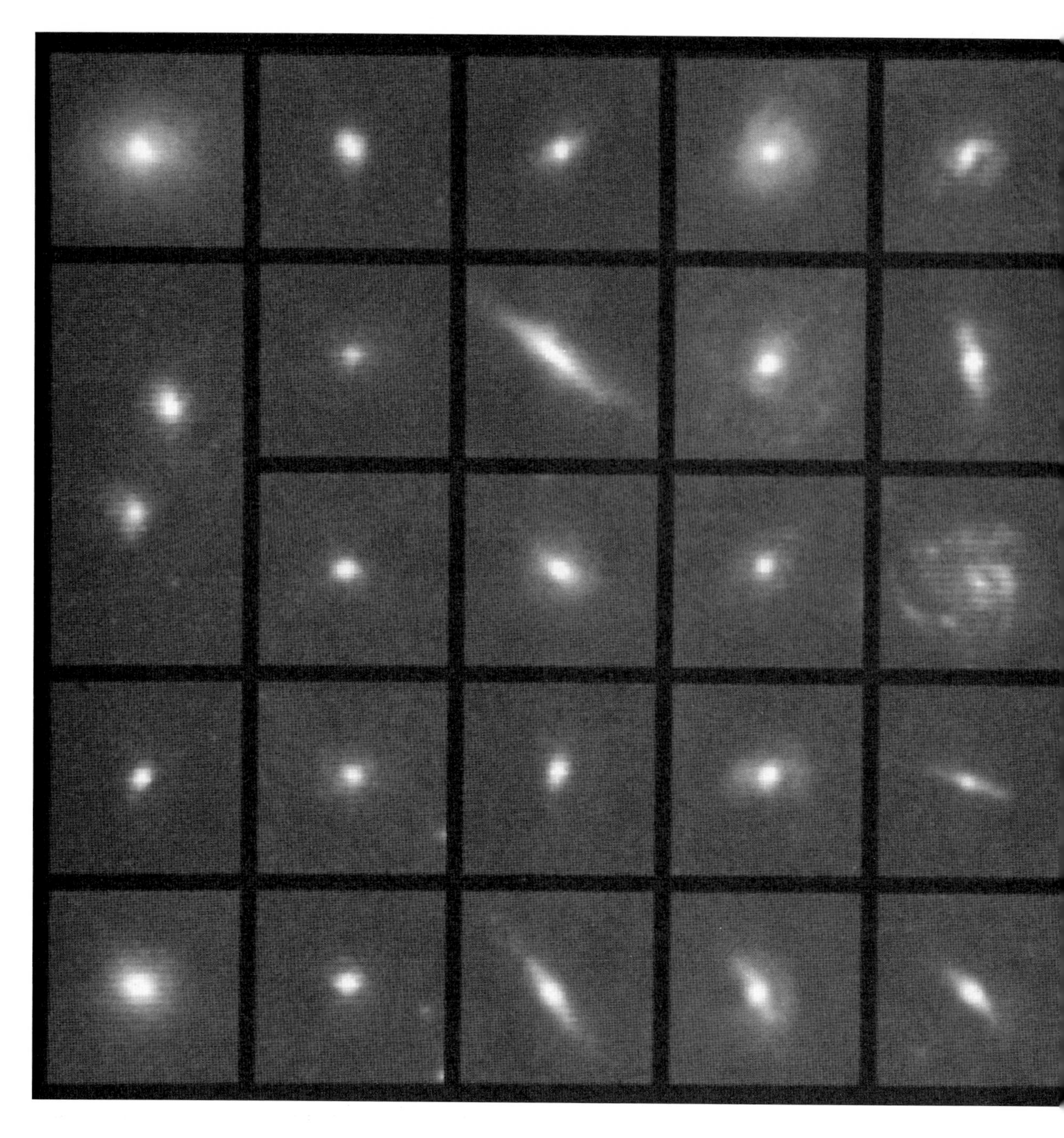

It was Edwin Hubble who first classified galaxies according to their morphology – their physical shape. Galaxies do not appear to be random agglomerations of stars but have structures which reflect the dynamics of stellar evolution: spirals are sites of vigorous starbirth whereas ellipticals tend to have older stars.

Credit: NASA

stars there differs from that of our own galaxy. The composition of the stars there is primordial and closer to that in the early universe.

A team led by Dr Nino Panagia of the European Space Agency has used Hubble to detect two dense clusters of stars in the Large Magellanic Cloud which appear to be linked, and where the deaths of

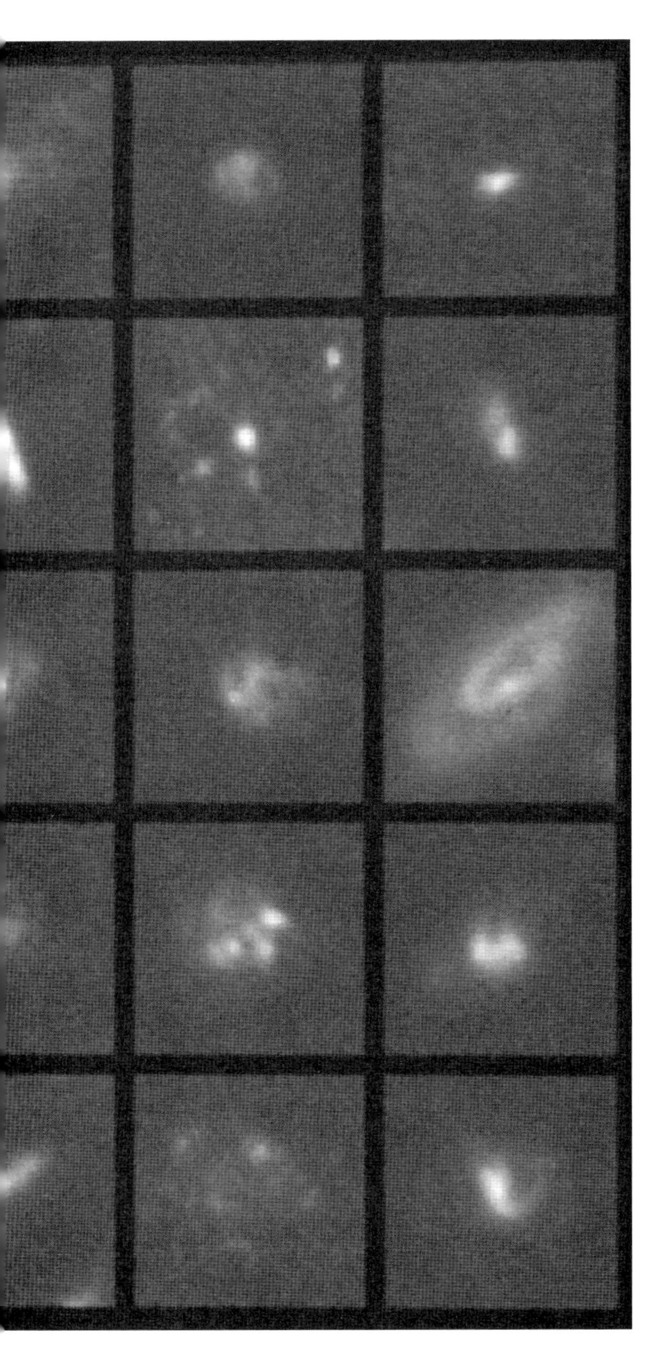

stars in one cluster are triggering the formation of new ones in the other. To do this they employed three characteristics of the Hubble Space Telescope: its ability to 'see' ultraviolet light; its light-gathering powers, which enable it to discern very faint stars; and its ability to focus sharply and tell distant objects apart, i.e. its high resolution.

They identified nearly 10,000 stars in one particular region of the Large Magellanic Cloud, where ground-based observations had previously succeeded in resolving only about 1,000 stars. Nearly two-thirds of the stars are about 50 million years old, but beyond them, about 200 light-years further away along the line of sight, lies a smaller cluster (about a fifth of the total number of stars) which are only 4 million years or so old. These stars are extremely hot and massive young stars. The presence of two distinct populations of stars so close together is unusual, and suggests that they might be connected through the process of stellar evolution. Hot gas from supernova explosions in the older cluster might have triggered the collapse of clouds of dust and gas into the formation of the newer stars. The suggested scenario is that an expanding shell of gas from the supernova explosions moved across space for 45 million years before hitting a wave of cooler dust and gas. The shock wave from this collision broke up the cloud, causing parts of it to contract under its own gravity and, eventually, to form stars. The new stars are themselves so massive that they will go into supernova phase in a few tens of millions of years, creating another shock wave of expanding hot gas and, perhaps, triggering further star formation in turn.

Hubble has been able to find evidence of starbirth in a more distant galaxy, known as M33, which lies just on the fringes of our 'local group'. M33 is a spiral galaxy, some 2.7 million light-years away – sixteen times as far away as the Large Magellanic Cloud. Despite its great distance, the largest ground-based telescopes can discern structure within the galaxy, including a nebula, designated as NGC 604, in an outer region of one of the spiral arms. Distance can be misleading; although this faint patch looks tiny even when viewed by one of the largest ground-based telescopes, Mount Palomar, it stretches nearly

Two star clusters 166,000 light-years away in the Large Magellanic Cloud (LMC), in the southern constellation Doradus. In a region 130 light-years across, the HST can identify more than twice as many stars as can be seen over the entire night sky with the naked eye. About 60 percent of the stars belong to the 50-million-year old dominant yellow cluster NGC 1850, while about a fifth of them are white, massive stars only 4 million years old and about 200 light years further away. Supernova explosions in the older may have triggered the birth of the younger cluster.

Credit: R. Gilmozzi (STScI/ESA), S. Ewald (JPL) and NASA

1,500 light-years across. Astronomers can see the nebula because it contains a huge area of dust and gas that is glowing, raising suspicions that this is an area where stars are being born.

Observations by the Hubble Space Telescope show that NGC 604 contains more than 200 young hot stars, each between fifteen and sixty times as massive as our own Sun. The dust and gas cloud is glowing because the intense radiation given out by

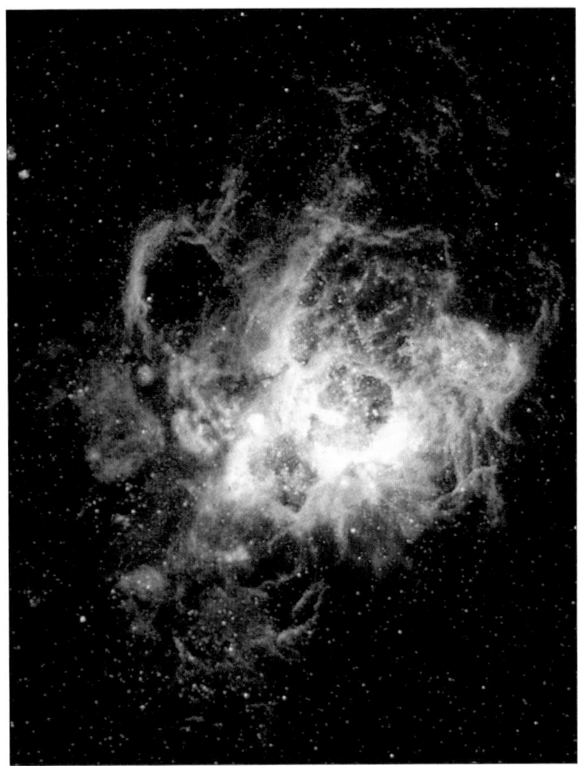

these incredibly active young stars is lighting it up from (as it were) the inside.

Hubble can still resolve details of starbirth in galaxies even further away, beyond our local group. In the centre of NGC 253, which lies some 8 million light-years distant, immense energies are being released in a violent burst of star formation. In fact, so much activity is going on in the heart of this spiral galaxy that it is considered to be what is known as a 'starburst galaxy'. Such galaxies emit a great deal of infra-red radiation, coming from the dust that has been heated up by the hot young stars. Hubble allows astronomers to peek into the middle of this galaxy and see the swirling patterns of dust and the gas glowing as they are excited by radiation from the new stars.

The effect of blast waves from supernovae – which was seen comparatively nearby in the Large Magellanic Cloud – can be seen at these sorts of distances also, at least by the Hubble Space

The arms of the spiral galaxies are the sites of stellar birth as here in one of the spiral arms of the galaxy M33. It is part of our 'local group' including the Milky Way and Andromeda but is on the outer fringes, some 2.7 million light-years away, in the constellation Triangulum. This particular region of intense starbirth activity, the nebula NGC 604, is unusually large, nearly 1,500 light-years across. The nebula is so big it can be seen by ground-based telescopic images (boxed on the left). The Hubble close up (right) gives much more detail on the dynamics of the gas clouds.

Credit: Hui Yang (University of Illinois) and NASA. Ground-based image: Palomar (Caltech) and AURA.

Telescope. In NGC 2366, an irregular galaxy which lies 10 million light-years away, Hubble has picked up a cloud of glowing gases which represent another star-forming region. The gas has been partially blown away by supernova explosions and by the wind from younger stars, leaving a 'hole' visible towards the top of the nebula.

But images from the Hubble Space Telescope have also led astronomers to believe that there

A visible light picture of the central region of a distant cluster of galaxies (CL 0939+4713). Because of its distance the cluster appears to us as it was when the universe was two-thirds its present age. It reveals the distribution of galactic shape in the remote past. Most of the spiral galaxies have odd shapes and suggest that the present-day spirals may have acquired the size and shape by a process of merger.

Credit: A. Dressler (Carnegie Institution) and NASA

can be other triggers for starbirth than 'just' the shock waves from exploding supernovae. One such trigger, they now think, is when two galaxies collide with each other.

The immensity of such an event can hardly be imagined; certainly ordinary words fail to capture the titanic forces at work. Because galaxies tend to congregate in clusters, and because they have been in existence for such a long time, there is a good chance that each will suffer at least one collision

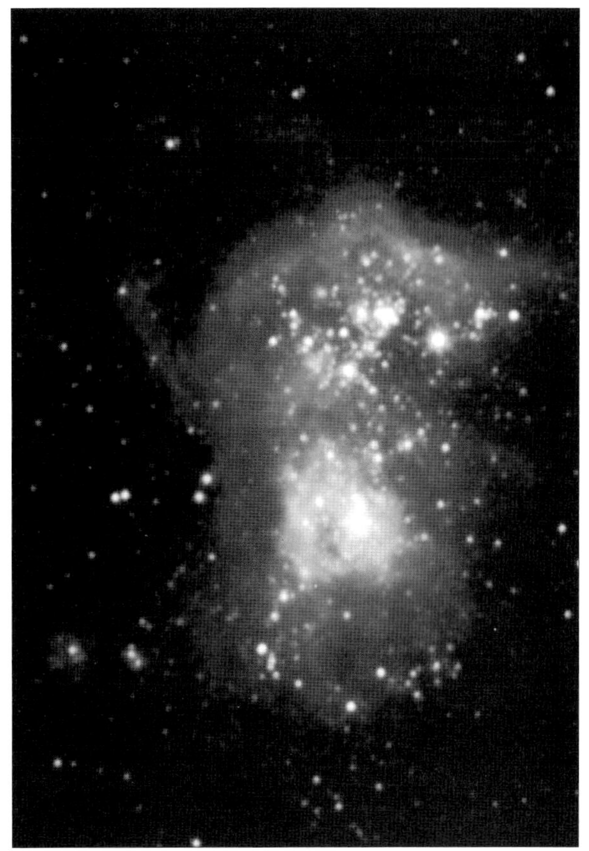

Left: A visible light image of the irregular galaxy NGC 2366 from the Canada-France-Hawaii (CFHT) 3.6-meter telescope on Mauna Kea, Hawaii. The white square shows a star-forming region (which has its own catalogue number, NGC 2363) and which Hubble's Wide Field and Planetary Camera 2 has been able to see in close-up (*right*). The brightest cluster of stars (2 million years old) is shrouded in the gas and dust from which it formed, whereas the 4-5 million year old cluster at the top of the nebula lies in a cavity blasted out by stellar winds and supernova explosions.

Credit: L. Drissen and Y. Dutil (CFHT) and NASA

during its lifetime, despite the immensity of the distances separating even those galaxies that are clustered together.

While the universe may seem a lonely place to us, on a galactic scale there is a certain amount of shoving and jostling for place. Even if individual stars do not actually bash into each other, the result of a near-collision between two galaxies will be immensely disrupting as the huge gravitational tides distort the galaxies' shape and structure. The shock wave generated by the encounter will bunch dust and gas together, thus creating perfect conditions for star formation. As we will see a little later in this chapter, they may also trigger conditions for a more spectacular star death than any yet encountered, in which stars are swallowed by the million to feed a supermassive black hole.

More than 250 million light-years away, a burst of star formation is the tell-tale signal that two spiral galaxies have run into each other. The Near Infra-Red Camera and Multi-Object Spectrometer (NICMOS) on board Hubble has snapped star formation in Arp 220 (i.e. the 220th object in the Atlas of Peculiar Galaxies compiled by the American astronomer, Halton Arp). The cores of the two galaxies, about 1,200 light-years apart, are

The starburst galaxy NGC 253, 8 million light-years distant, as seen (*left*) by the ground-based Lowell Observatory telescope. The Hubble image of the core (*right*) shows vigorous star formation going on in a region just 1,000 light-years across. The high rate of star birth was first identified by the infrared radiation from warm dust. Hubble reveals complex structures including luminous star clusters, 'lanes' of dense gas and dust, and streamers of hot gas. The existence of a bright, compact star cluster suggests that stars may often be born in dense clusters within starbursts, and that dense gas coexists with and obscures the starburst core.

Credit: Carnegie Institution of Washington, J. Gallagher (University of Wisconsin-Madison), A. Watson (Lowell Observatory), and NASA

orbiting each other. One of them appears to have the shape of a crescent moon, but is actually a spherical cluster of more than 1 billion stars, with the bottom part of the sphere concealed by a cloud of dust. The core of the other galaxy appears as a bright object to the left of the crescent.

Closer to our own local group, but still some 25 million light-years distant, the Hubble Space Telescope has captured in the picturesquely named Whirlpool galaxy the aftermath of an intergalactic collision that took place about 400 million years ago. The centre of this spiral galaxy has an intense concentration of stars, which are packed together 5,000 times more densely than in the neighbourhood around our own Sun. So densely, in fact, that if there were any observers on planets in the middle of the galaxy, the sky will be permanently bright with their radiance. This intense centre is only about 400 million years old, and is surrounded by a belt of stars that are at least 8 billion years old. Further out still lie clusters of even younger stars, perhaps only 10 million years old. They are much closer to the core of the galaxy than would be expected under normal circumstances: such young stars should be found far out in the spiral arms.

The answer to this conundrum seems to be found in the fact that the Whirlpool galaxy has a dwarf companion galaxy, which is so close that it appears almost to be connected to it by an extension to one of the spiral arms. Astronomers believe that the dwarf galaxy passed so close to the centre of the Whirlpool 400 million years ago as to stir up clouds of gas and dust, which triggered the outburst of star formation in the centre. The wrestling

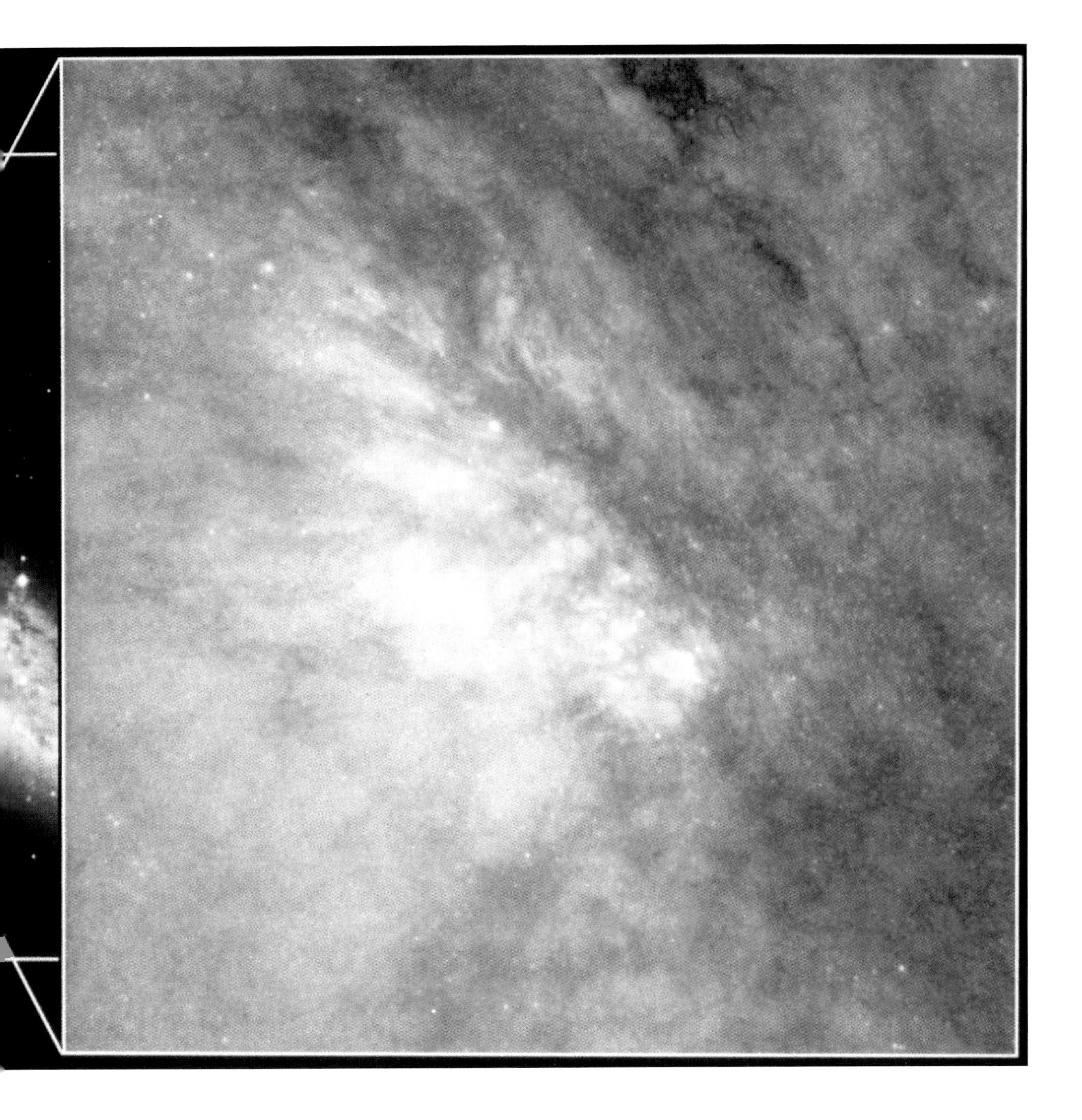

between the two galaxies is by no means done, and the outer clusters of younger stars near the core may be a recent consequence of their interaction.

Perhaps the most spectacular galactic collision seen to date is that of the pair of galaxies known as the Antennae galaxies. These get their name from the long trails of stars and other debris stretching out as a result of their collision, which have fancifully been likened to an insect's antennae. The Hubble Space Telescope discovered evidence that the collision has triggered the formation of more than 1,000 clusters of bright young stars, by squeezing and compressing the giant molecular clouds within the galaxies.

A collision between two spiral galaxies in Arp 220 has provoked a burst of star formation, captured by the Near Infra-red Camera and Multi-Object Spectrometer (NICMOS). The object shaped like a crescent moon is a cluster of 1 billion stars – the remains of the core of one of the colliding galaxies. Its lower half is obscured by a disk of dust about 300 light-years across, which may be swirling around a black hole. The core of the other galaxy is the bright round object to the left. About 1,200 light-years apart, they are orbiting each other.

Credit: R. Thompson, M. Rieke, G. Schneider (University of Arizona), N. Scoville (CalTech) and NASA

These mighty galactic pile-ups create casualties as well as new stars. Hubble has discovered many stars that are flung out of colliding galaxies, ending up wandering space in permanent exile.

In the Virgo cluster of galaxies, for example, it is estimated that there could be as many as 1 million million stars wandering through the void of intergalactic space, ripped out of their home galaxies as the result of a collision. Early in 1997 an international group of astronomers from Canada, the UK

About 400 million years ago, a dwarf companion galaxy passed close to the Whirlpool galaxy (M51), a spiral some 23 million light-years from Earth in the constellation Canes Venatici (the Hunting Dogs). The close encounter stirred up dust and material for new starbirth in the central region and is believed to be responsible also for unusually vigorous star formation in a bright ring of young stars further out.

Credit: N. Panagia (STScI and ESA) and NASA

and the USA announced that they had found more than 600 exiled stars in just one small patch of the Virgo cluster. All of them appear to be red giants, stars flaring up before the twilight of their lives. Red giants are sufficiently bright that they can be seen over the 60 million or so light-years that separate Earth from the Virgo cluster. There may well be many more stars – perhaps as many as 10

Above: The massive centre of M51, the Whirlpool galaxy. The concentration of stars at the centre is about 5,000 times higher than in the Milky Way galaxy. At the very core may lie a black hole that is producing powerful radio jets.

Credit: N. Panagia (STScl and ESA) and NASA

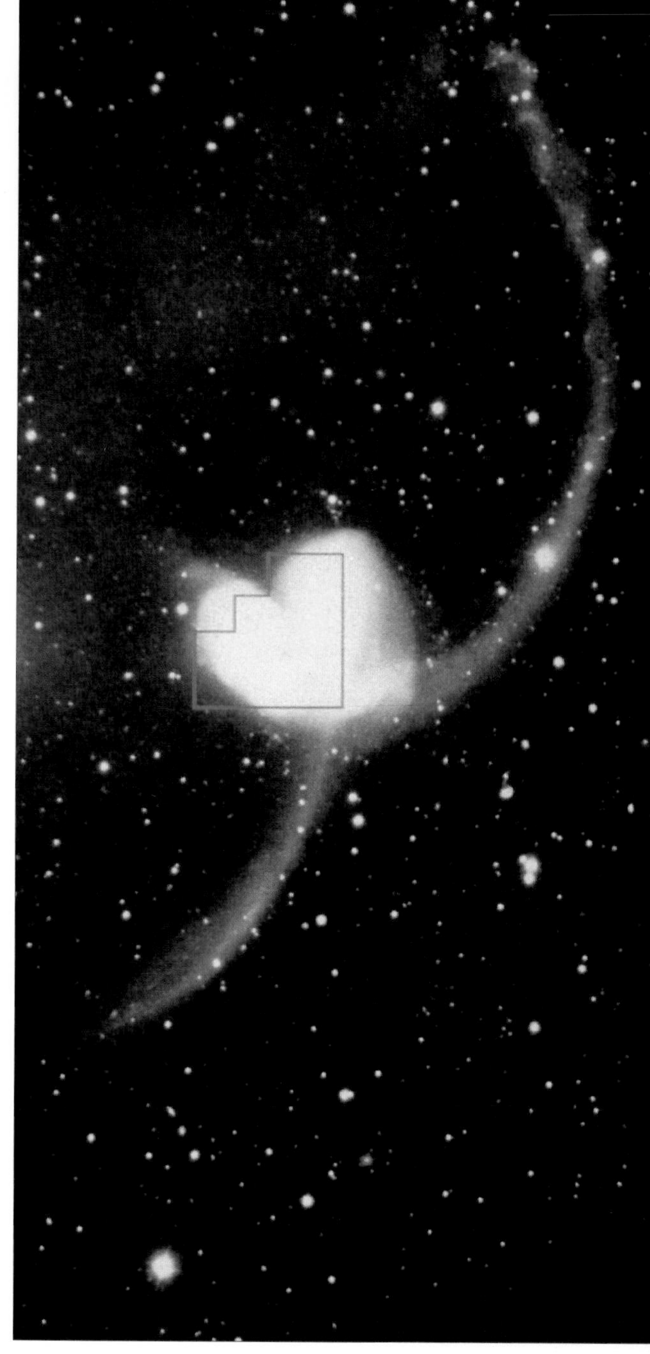

Right: More than 1,000 bright, clusters of young stars have been kicked into life as a result of the head-on collision of these two galaxies. On the left are the Antennae galaxies (NGC 4038 and NGC 4039) as seen through a ground-based telescope. The galaxies are located 63 million light-years away in the southern constellation Corvus. In the close-up on the right, the cores of the two galaxies can be seen glowing orange and are criss-crossed by streamers dark dust. The dark region in the centre is the overlap – a wide band of dust. The sweeping spirals, traced by bright blue star clusters, shows the star formation triggered by the collision.

Credit: B. Whitmore (STScI), and NASA

million – that are too faint to be seen at that distance, even by the Hubble Space Telescope. By extrapolating from the ones they can see in a small region to the whole volume of the Virgo cluster, astronomers have come up with the estimate that at least 10 per cent of the all material contained in the cluster may be accounted for by exiled stars.

If any of these stars were to have planets and intelligent life then the night sky would appear very different from what we can see from Earth. No stars

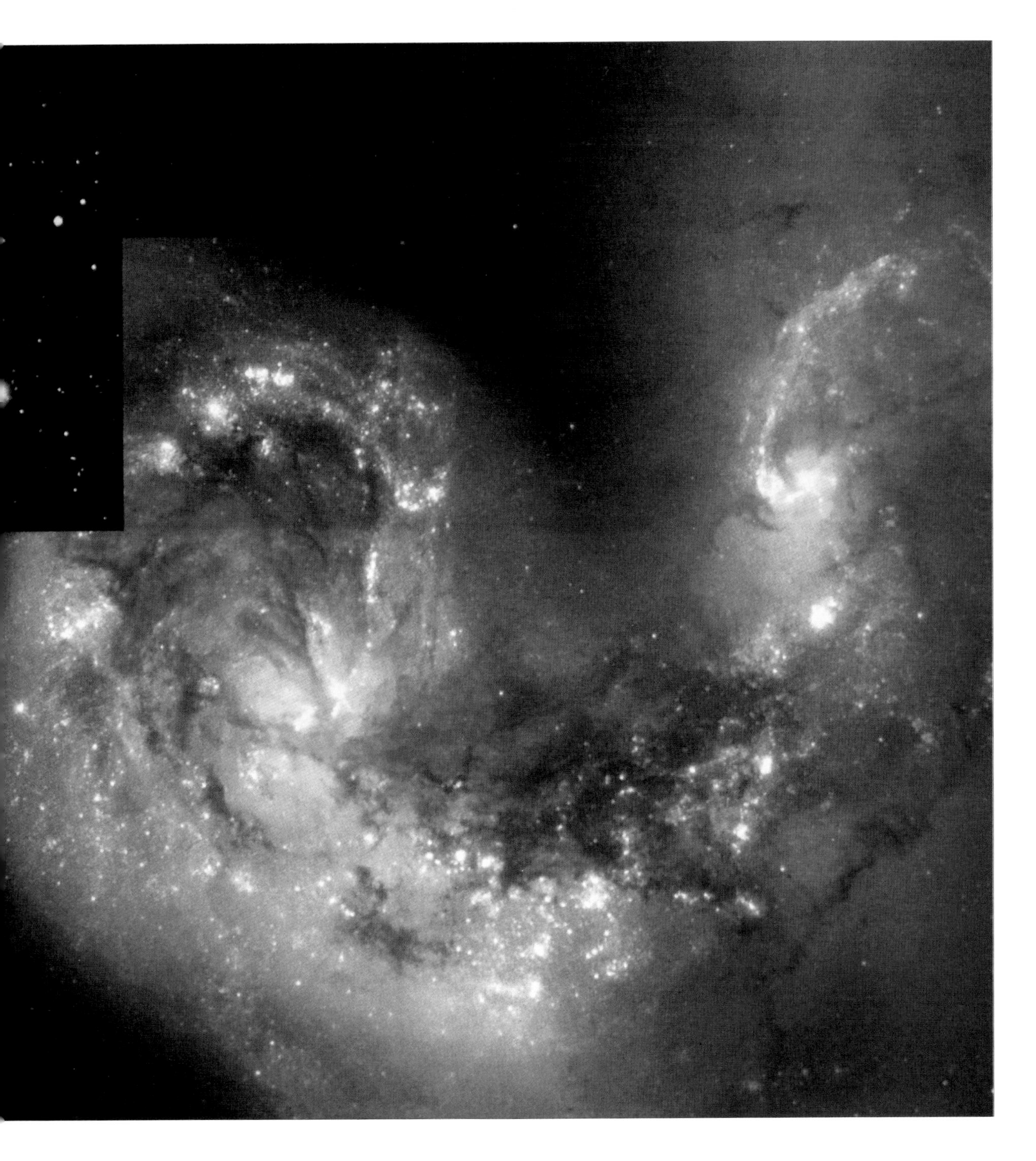

would be visible once the system's own sun had set; individual stars would be too far away to be seen. The only astronomical objects that would be visible would be the fuzzy patches of light representing distant galaxies, perhaps including the one from which the star and its planets had been ejected. But these stars are so far from any one galaxy that their motion through space would be determined by the gravity of the cluster as a whole rather than by the pull of any one galaxy. This is true isolation indeed.

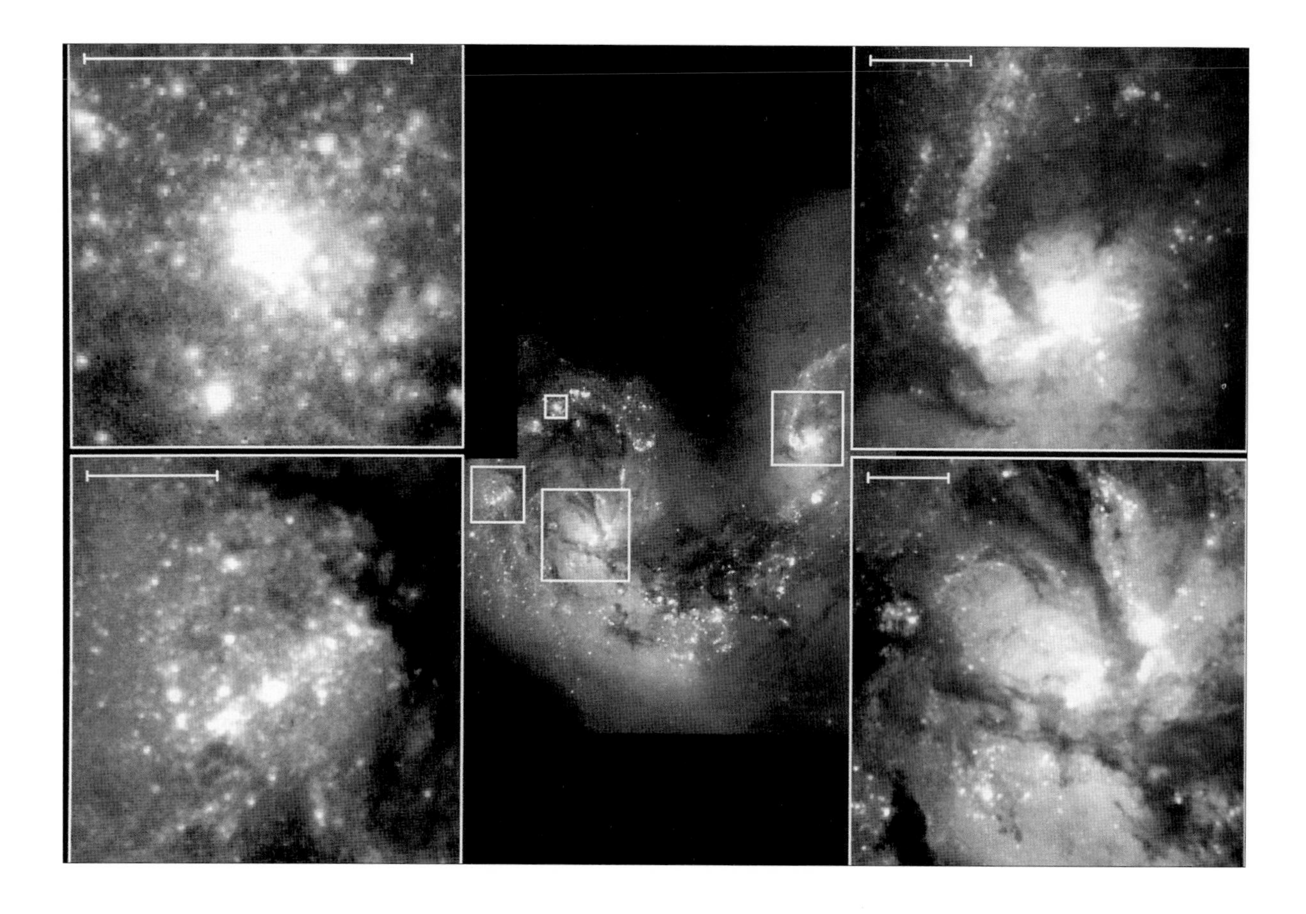

Left: The collision of the Antennae galaxies kicks new stars into life in these brilliant clusters, of which the brightest may contains as many as a million stars. The very young stars are only a few million years old.
Right: Dust and gas has been trapped in the centres, each galaxy dimming and reddening the starlight by scattering the shorter, bluer wavelengths.
This and the previous natural-colour image are each composites of four separately filtered images taken with the Wide Field Planetary Camera 2.

Credit: B. Whitmore (STScI), and NASA

Astronomers had predicted that one consequence of intergalactic collisions would be the creation of wandering, outcast stars. As Virgo is the nearest large cluster of galaxies, ground-based telescopes have been examining it for signs of the outcasts. Astronomers expected to find a diffuse halo of light in the cluster but, as so often in galactic astronomy, the effects of the Earth's own atmospheric glow bedevilled attempts to obtain an unambiguous observation. Now Hubble has shown not only that such stars do exist as predicted, but also that they are to be found stranded at least 300,000 light-years from the nearest big galaxy. This is more than three times the diameter of the entire Milky Way.

Work done by Hubble suggests that the collision of two galaxies could also fire into life the most mysterious and awesomely powerful objects in the universe: quasars.

Quasars are the most luminous objects known, some of them shining with a power 1,000 times greater than that produced by an entire galaxy of stars. But all that energy – equivalent to more than 100,000,000,000 stars – is coming from a single object not much larger than our own solar system. There is only one process known that can produce

The Hubble Space Telescope's 100,000th exposure, in June 1996, was of a quasar about 9 billion light-years from Earth. The quasar is the bright object in the centre, with a slightly fainter elliptical galaxy just above it. The galaxy is almost directly in front of the quasar and 2 billion light years closer to Earth.

Credit: C. Steidel (CalTech and NASA)

such energy, and it is a massive black hole surrounded by a swirling disk of gas. The disk represents the debris of entire stars that have been torn apart by the tidal forces of the black hole's gravitation. The black hole is swallowing this gas at a rate equivalent to one Sun a year.

Quasars are the ultimate sharks of the intergalactic medium. The discovery and analysis of quasars is a story of astronomers using all the instruments at their disposal to glean scraps and hints about the immense forces at work. Even so, quasars remain enigmatic. They were identified in the early days of radio astronomy as sources of intense radio emission, but it was not until 1963 that astronomers managed to track down the optical counterpart to one such source. It was the 273rd object listed in the third Cambridge catalogue of radio sources, and is therefore known as 3C 273. When the optical counterpart was located, it proved to be a faint star-like object that appeared to be immensely distant. Here were two puzzles: this quasar was shining hundreds of times more brightly than whole galaxies that were known to be roughly the same distance away, yet all this radiation appeared to be coming from a single, compact, star-like source. More examples were discovered and, unable to fit them into existing categories, astronomers invented a new name for them: Quasi-Stellar Radio Sources or quasars. (Eventually, it

The quasar PKS 2349 is 1.5 billion light-years away and in the act of absorbing a companion galaxy. *Above:* the distorted shape and long, thin tidal arms suggest a galactic collision. (The thick bright line above it is just a background galaxy seen edge on.) *Opposite:* the same image, at different contrast, shows more clearly the companion just above the main galaxy. In about 10 million years the black hole which powers the quasar will consume the companion galaxy.

Credit: J. Bahcall (IAS, Princeton) and NASA

turned out that even this was a misnomer: few quasars emit intense radio waves, so the designation 'quasar' is now taken to stand simply for 'quasi-stellar object'.)

Every detail about quasars is perplexing. For a time, some astronomers refused to believe that these objects could possibly be as far away as the measurements suggested. The brightness of an object as measured from Earth depends on how

intrinsically luminous the object is in itself and on how far away it is. If two objects have the same intrinsic luminosity but one is twice the distance of the other from Earth, then it will appear to be only a quarter as bright when observed from Earth. A source three times as far would appear one ninth as bright. Thus if quasars were (say) 1 million times closer than had been thought, then it would be easier to explain how so much luminosity could come from such a small object. However, the idea that quasars might not be so far away has been dropped in the face of several pieces of evidence, not least the discovery that distant but ordinary galaxies lie between us and them.

It is the smallness of the region from which all this energy radiates out that is the second puzzle presented by quasars. It is bad enough that they shine with the luminosity of 100 (or, for the really

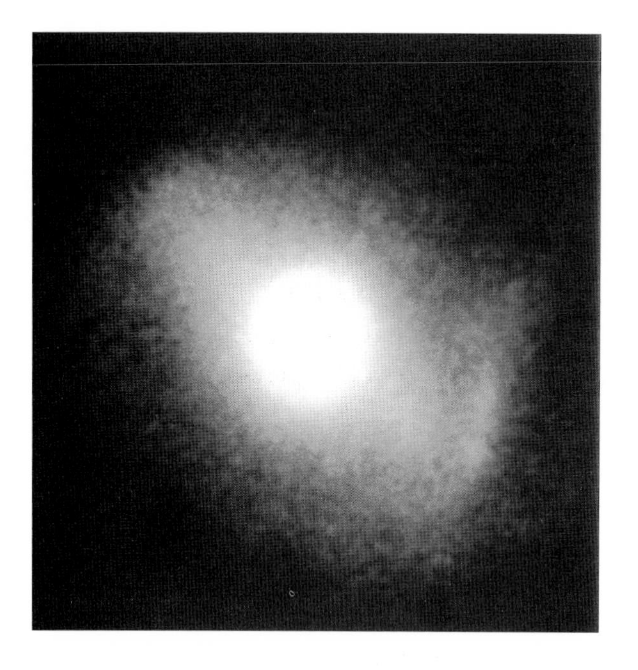

Quasar 1229+204 lies in the centre of a galaxy that is colliding with another fuelling not only the massive black hole at the centre but also triggering new star formation. The ground-based Canada-France-Hawaii Telescope first identified the barred spiral galaxy in 1229+204, but Hubble reveals more details, including a string of knots, which are probably clusters of massive young stars.

Credit: J. Hutchings (Dominion Astrophysical Observatory) and NASA

distant ones, 1,000) galaxies, but their brightness varies rapidly on the scale of a couple of days or less. The rapidity of the variation puts limits on how big the object can be. If it varies in a day, virtually all the object's light or radio wave output must be changing over a period of less than a day. This in turn means that the object itself can be no larger than one 'light-day' in radius (a paltry 16 billion miles/26 billion kilometres or so) for otherwise the brightening would take longer than one day, as the light from the edges of the object would take a day longer to reach us than light from the front of the quasar. A light-day is about 170 times the distance from the Sun to the Earth, or only about twice the diameter of Pluto's orbit around the Sun.

The only explanation that astronomers can come up with is that the luminosity of quasars is due to supermassive black holes lurking at their core, swallowing dust, gas and stars. Although the term black hole was invented as recently as 1967 by the American physicist John Wheeler, the concept dates back to 1784. An obscure English clergyman, the Reverend John Michell, speculated then that it might be possible for a star to be so big and so massive that its gravity prevented even light from escaping. The concept was dreamed up again, quite independently, a decade later when the French mathematician Pierre-Simon de Laplace speculated about the possibility of 'dark stars'. But it was only with Einstein's General Theory of Relativity, published in 1915, that the idea was put on a sound theoretical footing. The following year, the German mathematician Karl Schwarzschild produced calculations showing how the gravity of a star could 'bend' space in its vicinity, and how a sufficiently massive star would have a gravitational field so strong that nothing – not even light – could get out.

It seems likely that some black holes arise as the remnants of massive stars after they have gone supernova. Their cores might be so large that instead of collapsing to the great density of a neutron star, they would pass through the neutron phase and contract even further, disappearing out of existence as a black hole. The intense gravitational force of the black hole would force material – dust and gas – in the vicinity to move towards it, swirling round and round like the water going down a plughole, before the dust and gas fall into the hole and out of existence completely. Extinction in a black hole is a far more efficient way of converting hydrogen gas into energy than 'burning' it in the centre of a star to form helium by nuclear fusion. Here then is one way in which a quasar might shine so bright: the black hole at its centre can convert matter into energy more efficiently than a normal star. However, there is one important difference between the black hole that results from a stellar collapse and the one that might power a quasar. The black hole at the centre of a quasar would have to be millions of times more massive to achieve the power outputs that astronomers have measured.

There is a second way in which the material in the disk of matter swirling around a black hole can radiate energy: by harnessing some of the energy with which the black hole is spinning. Here on Earth, the brakes in a car will get hot as they are applied and slow the rotation of the wheels; they convert the energy of the motion of the wheels into heat energy. In the disk round a black hole that is the powerhouse of a quasar, the principle is the same: the material in the disk of matter acts as a brake to slow the spinning of the black hole, and by doing so extracts energy which is eventually radiated away into space. Calculations that this radiation should be particularly evident in the ultraviolet part of the spectrum have been confirmed by observations.

But if the process is reasonable, the quantities involved are immense. Given the luminosity of a quasar, an enormous amount of matter must be being converted into energy. Black holes offer a solution to this problem too, for they can eat up entire stars at a time, and do so often. Calculations suggest that the most luminous quasars would need to have black holes as massive as 100 million Suns, and to be consuming stars at the rate of about one a year. In densely packed galaxies it is entirely possible that dust and gas and stars could fall into a supermassive black hole at this sort of rate. The black hole itself would have to be around 0.1 per cent of the mass of the entire galaxy. Each time it swallowed a star, of course, its mass would increase. This suggests that the black hole could have started small and built up, over the lifetime of the galaxy, to such a massive presence.

One of the problems in studying quasars is that they are so incredibly remote from us that it is difficult to find 'normal' objects at that distance with which to compare them. Galaxies at the same distance as the quasar shine far less brightly, and are extremely difficult for ground-based telescopes to make out against the background glow of the Earth's atmosphere. This difficulty is compounded if astronomers want to try to study the galaxy within which the putative black hole sits: the light from the galaxy (if it exists at all) will be blotted out by the fantastic glare from the quasar at its heart.

So it is not surprising that scientists have used Hubble to cast some light (so to speak) on the quasar enigma.

Mike Disney, the leader of one such team of astronomers, from the University College of Wales at Cardiff, remarked: 'I gave up on studying quasars twenty years ago because I realised we had to wait for a space telescope to provide a clear enough view for solving the mysteries.'

Professor Disney's team found that the collision between two galaxies appears to be the trigger

Homes of the quasars. *Clockwise from top left:* Quasar PG 0052+251 at the core of a normal spiral galaxy; Two galaxies, colliding at 1 million mph, may be fuelling quasar IRAS04505-2958; Quasar 0316-346's dust trail suggests that its galaxy is still reeling from a collision long ago with a passing galaxy; Two galaxies may have orbited each other several times before merging, leaving loops of glowing gas around quasar IRAS13218+0552; Quasar PG 1012+008 merges with a bright galaxy; Quasar PHL 909 inhabits an elliptical galaxy.

Credits: J. Bahcall (IAS, Princeton) M. Disney (University of Wales) and NASA

mechanism for firing up a quasar into action. 'In nearly every quasar we look at, we clearly see one galaxy apparently swallowing another,' Professor Disney stated.

His group has looked at quasars that are emitting strongly in the infra-red spectrum, suggesting that they were located within spiral galaxies (the infra-red emissions tend to be a sign that dust and gas are present, and, as we have seen, such clouds of matter are typical of the arms of spiral galaxies). According to Professor Disney, the Hubble images reveal galactic collisions 'where two giant spiral galaxies like our own Milky Way have crashed head-on into one another, and flung off pieces violently in all directions. Some of those bits seem to have finished up in the nucleus of one of the spirals where there is probably a giant black hole feeding on it.'

The implication of this work is that a galaxy may well have contained at least one small black hole – probably the collapsed remnant of a supernova star. Such a black hole would be reasonably quiescent unless its environment was stirred up, pushing dust and gas into its gravitational ambit. Then the black hole would grow with every bit of material that fell into it. The immense tidal effects of a collision between two galaxies would, of course, provide just such a disturbance.

But, as so often, what the Hubble Telescope appears to give with one set of observations, it takes away with another. A different team of researchers, led by Professor John Bahcall of the Institute for Advanced Study at Princeton, conducted a survey of a score of quasars and found that about half of them sit in host galaxies that appear to be undisturbed, and certainly not interacting or colliding with other galaxies.

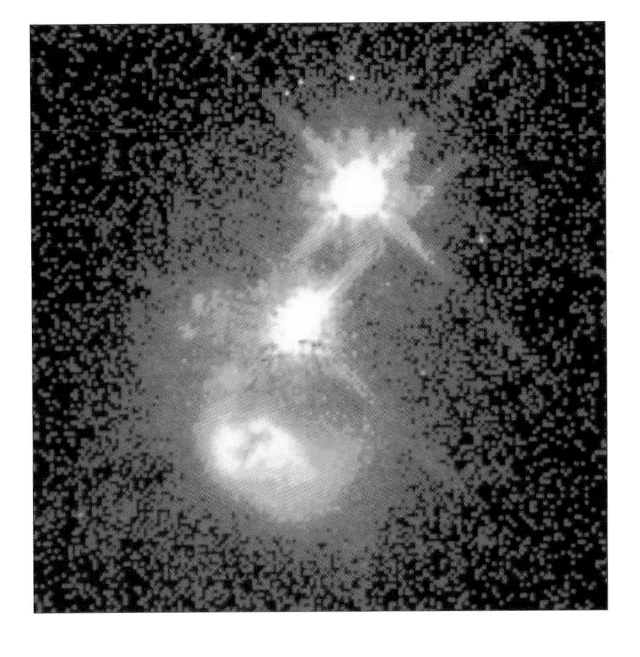

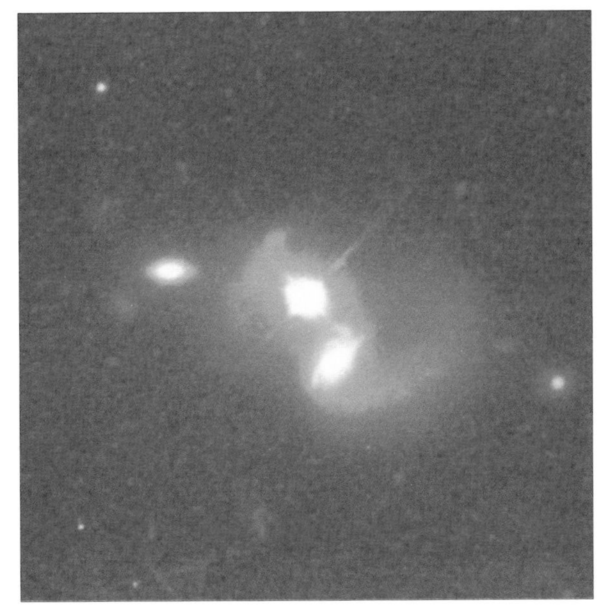

Professor Bahcall concludes that there may be several different ways of switching a quasar on, some of them perhaps more subtle than astronomers have yet thought of. 'Some of the galaxies we observed don't appear to know they have a quasar in their cores,' he commented.

The fact that some galaxies appear relatively unperturbed could mean that quasars are relatively short-lived and that many galaxies – including possibly our own Milky Way – might have experienced quasars a long time ago. This possibility is enhanced by one outstanding feature of quasars: the further away astronomers look, the more quasars they find.

One of the quirks of observing great distances in astronomy is that a telescope is also acting as a time machine. Although a telescope may be taking

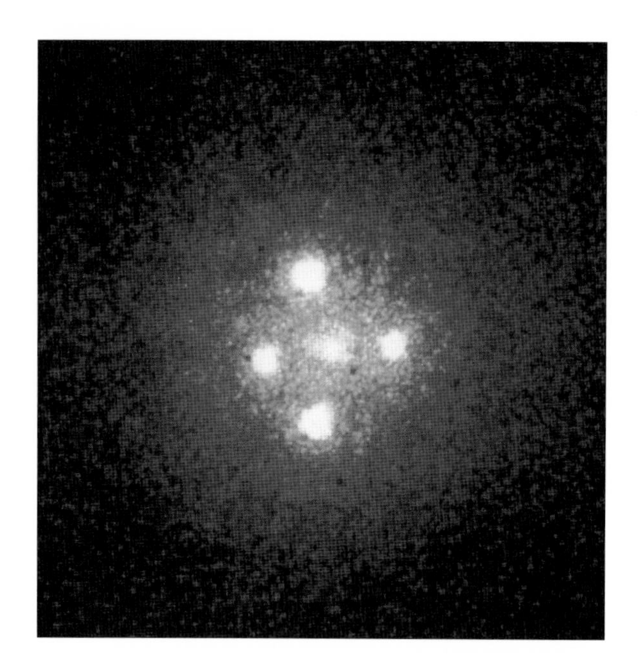

Cross-shaped, gravitational lenses such as this might provide astronomers with a powerful way to probe even further into the universe. The enormous gravitational field of a massive object, such as a galaxy, can bend light so as to magnify, brighten (but also to distort) the image of a more distant object.

Credit: NASA

pictures now, it is not picturing the state of the universe as it is right now, but as it was many millions (if not billions) of years ago. In the case of the quasars, all the observations are of unbelievably distant objects, and therefore objects as they were when the universe was very young, possibly just a few billion years old. It is possible that all galaxies went through a 'quasar' phase when they were young, and that when eventually the amount of material falling into the black hole at the quasar's heart decreased, so the stupendous outpourings of energy ceased; in effect, the quasar's light would be switched off.

This (inevitably) raises yet more questions, this time about the evolution of galaxies. Were black holes there before the galaxies? And did they 'seed' the formation of a galaxy around them, either drawing in gas and dust, which would eventually clump together to form stars, or drawing together already formed stars to create celestial star cities? Or did the galaxies precede the formation of their black holes? Perhaps, astronomers speculate, the centres of galaxies were so densely packed that massive stars collided with each other until they could not sustain their own weight and collapsed to become a single black hole.

And yet another question: what happens to the black hole when the fuel runs out? It will not disappear, but will be sitting there quietly, almost invisible, waiting until (say) a galactic collision stirs up the dust and starts feeding its voracious appetite once again. This speculation could have interesting implications for the future of humanity; our galaxy is a pretty quiet sort of place, but if there is a dormant black hole at its centre, any turbulence could fire it into action, and our entire galaxy would be bathed again in the energy pouring out from a central quasar.

While quasars are by and large immensely distant from us, there are other, closer galaxies which also appear to have highly active centres. While they are not anything like as powerful as quasars, their overall character appears to have many factors in common with them. One group are known as Seyfert galaxies, after the American astronomer Carl Seyfert who started his definitive studies of these objects in 1943. A second group was discovered in the 1950s with the advent of radio astronomy. They appear to channel most of their energy output into radio waves, rather than visible light, ultraviolet or X-rays. These 'radio galaxies' emit energy in the form of radio waves at rates about 1 million times greater than the radio 'noise' from galaxies such as our own. Astronomers now lump all these types into one category – 'active galaxies' – and assume that all of them have a supermassive black hole at their centres feeding voraciously on stars and gas and dust.

In 1994 the Hubble Space Telescope demonstrated just how big 'supermassive' can really be when it detected a black hole weighing up to 2,400,000,000 (2 billion, 400 million) Suns in the nucleus of the active galaxy M87. Compared to quasars, which are thousands of millions of light-years distant, M87 is nearby: 50 million light-years away in the direction of the constellation Virgo. It is a prominent radio source, although the power level is way below that produced by a quasar. It also has a bright optical 'jet' of material emanating from the centre. In the late 1970s astronomers from the California Institute of Technology claimed that they had obtained evidence from ground-based observations of a vast black hole at the centre of this galaxy. They claimed to have detected two tell-tale signs of a black hole: an enhanced concentration of stars in the vicinity of the area suspected of containing a black hole; and an increase in the speeds of the stars the closer they are to that area. But their observations, and the interpretation put upon them, were hotly debated over the following years.

The advent of the Hubble Space Telescope has definitely settled the matter. It produced sharp pictures clearly showing the congregation of stars, and revealed a disk of glowing gas around a central point – the black hole. A straightforward picture of the nucleus of M87, taken with the Wide Field and Planetary Camera, does not look particularly spectacular. None the less, the team of astronomers were surprised to get such clear evidence of the very existence of the disk of gas. But it was when they employed Hubble's Faint Object Spectrometer as a celestial speedometer – to measure the speeds of the gas on either side of the disk – that they got the definitive evidence. The light from the part of the disk that is moving away from us in its orbit is shifted to the red end of the spectrum, whereas the gas on the other side, coming towards us on this bit of its orbit, is blue-shifted in its spectrum. These observations yielded the startling result that at about sixty light-years from the black hole, the disk of gas is rotating at speeds of around 470 miles (750 kilometres) a second. (For comparison, the Earth, which is eight light-minutes from the Sun, orbits at a mere 19 miles/30 kilometres a second. The more distant the orbit, the slower the velocity: Pluto orbits at just under 3 miles/5 kilometres a second.) From the speed at which the gas is rotating, it is comparatively easy for astronomers to work out the mass of the central object: 2.4 billion times as massive as the Sun. Nothing other than a black hole could possibly pack so much mass into such a small space. 'If it's not a black hole, it must be something even harder to understand,' commented one of the astronomers, Dr Richard Harms.

After this, observations of other supermassive black holes came swiftly. Even the Hubble Space Telescope can find only a few gas disks whose rotation can be measured sufficiently well to ascertain whether a black hole lurks at the centre, and so scientists used the Faint Object Spectrograph to measure the motions of stars in the centres of the galaxies as they circle around black holes. In 1996 it was announced that the technique had paid off with the detection of a supermassive black hole at the centre of the galaxy NGC 3115. This lies some 30 million light-years away in the constellation of Sextans. The tell-tale motions of the stars reveal that the black hole must have a mass of around 2 billion times that of our own Sun.

It sounds a simple technique, in principle. It is, after all, a method for detecting black holes by their most obvious characteristic: the gravitational pull they exert on objects in their surroundings. But it is easy to jump to conclusions; even without the presence of a black hole, stars in the centres of galaxies often move about very rapidly and in

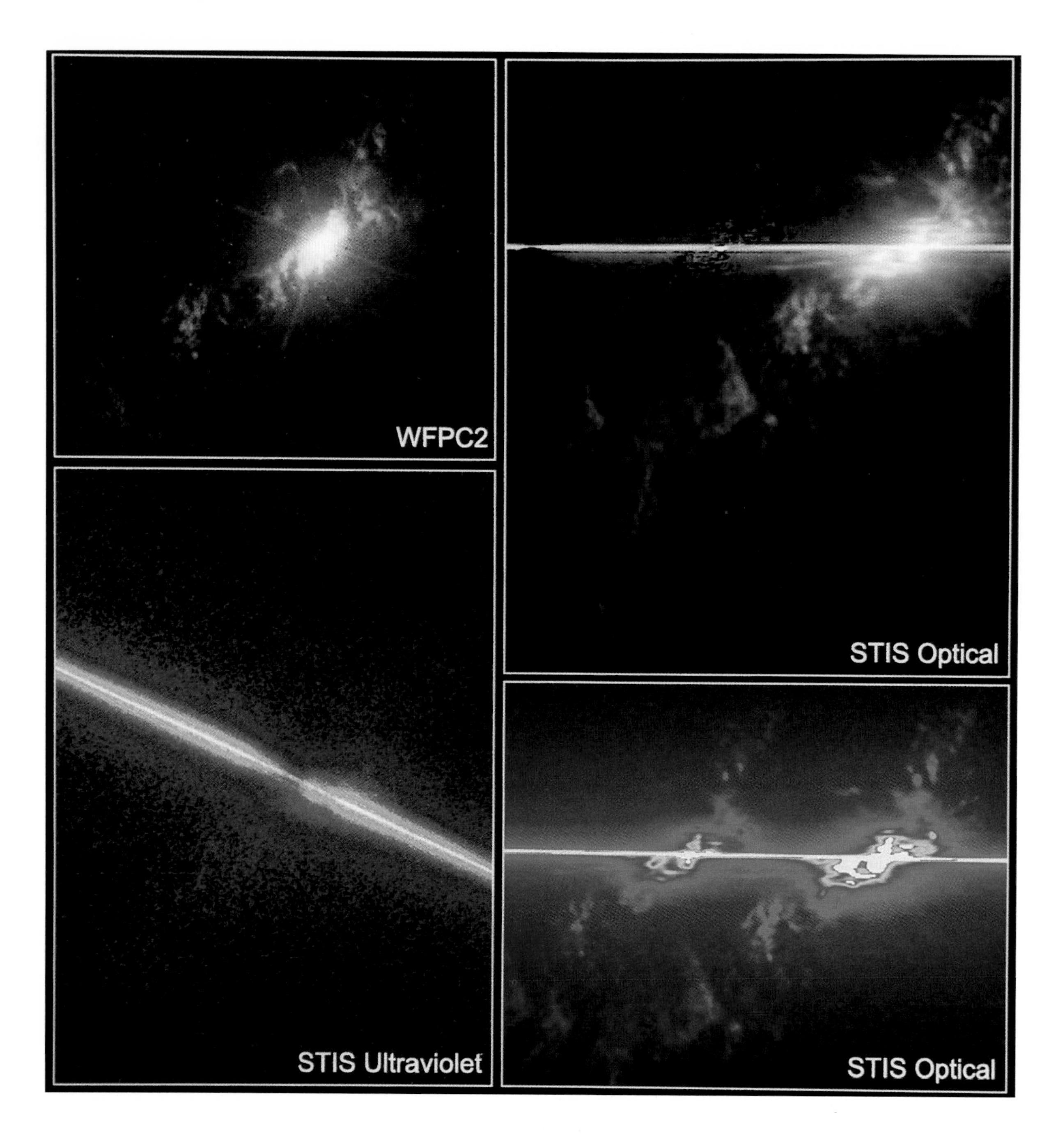

Detailed dynamics of Seyfert Galaxy NGC 4151, revealed by the Space Telescope Imaging Spectrograph (STIS). *Top left:* an image taken at a wavelength to show oxygen gas. The two 'STIS optical' images are spectra revealing that, of the two cones of gas emission (powered by a supermassive black hole), one is coming towards us, while the other is receding. The UV spectrum is carbon emission from closer to the core.

Credit: J. Hutchings (DAO), B. Woodgate (GSFC), and NASA

random directions. Their observed motion can therefore be complicated, and needs to be carefully dissected to ensure that the effect – an apparent attraction by a black hole – is real and not just some bizarre piece of interstellar dynamics.

The fine-tuning of the technique that tracked down the black hole hiding in NGC 3115 opened

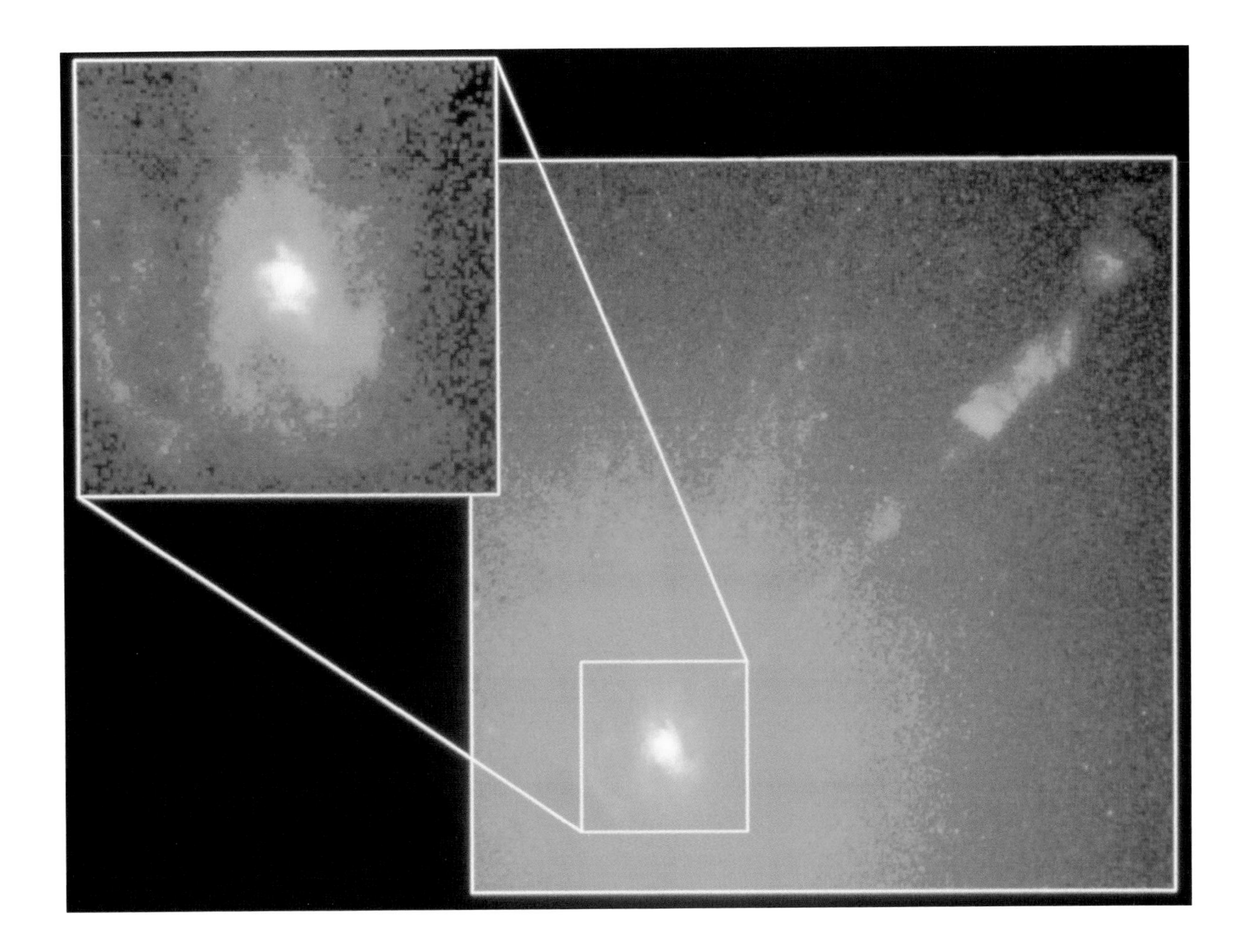

Measurements of the velocity of the spiral-shaped disk of hot gas surrounding the core of active galaxy M87 show that the disk is rotating so rapidly the galaxy must contain a massive black hole at its hub. M87 lies 50 million light-years away in the constellation Virgo and the object at its centre weighs as much as 3 billion Suns, but is concentrated into a space no larger than our solar system.

Credit: H. Ford (STScI/Johns Hopkins) et al and NASA

the way to a systematic survey to try to find evidence for massive black holes in ordinary galaxies, as well as in Seyfert and other active galaxies. Within a year, Hubble observations had led a large team of astronomers to the astonishing conclusion that virtually every large galaxy has a supermassive black hole lurking at its nucleus. Its leader, Douglas Richstone of the University of Michigan, concluded that these black holes are, in effect, 'fossil' quasars. Although quasars are comparatively rare nowadays – there is only one quasar for every 100,000 galaxies – it appears from these results that they were once common, and that the black holes which powered them are dormant but still around.

One might imagine black holes as stars with eating disorders. Most supermassive black holes today are anorexic, eating virtually nothing at all, and therefore detectable only indirectly by virtue of their immense gravity, which influences the orbits of stars in their vicinity. But a few black holes, when they get the opportunity, go bulimic, bingeing on all the stuff they can cram in and, in the process, spewing out immense quantities of radiation in the form of radio galaxies, Seyfert galaxies, or quasars.

The team found evidence for black holes in three normal galaxies. Two of the galaxies belong to a group relatively close to us, the Leo Spur, about 32 million light-years away in the direction of the Virgo cluster. One black hole, in galaxy M105, has a mass of about 50 million solar masses, and the other is twice as massive. The third galaxy is slightly more distant, about 50 million light-years away in the Virgo cluster itself. This galaxy, NGC 4486B, houses a 500 million solar mass black hole and is a small satellite of M87, whose 2.4 billion solar mass black hole was established by the Hubble Space Telescope in 1994.

In February 1997 Hubble's ability to detect black holes was boosted when as part of the second servicing mission (carried out by astronauts) the Faint Object Spectrograph was removed and replaced with an even more sensitive instrument, the Space Telescope Imaging Spectrograph (STIS). This machine, it was calculated, ought to be about forty times as quick in detecting supermassive black holes. And indeed, within the space of a few weeks it revealed yet another massive black hole, at the centre of M84, another member of the Virgo cluster of galaxies.

The fact that supermassive black holes might be common is fascinating enough. But one aspect of the Hubble observations that is particularly intriguing astronomers is the indication that the mass of the black hole may be related to that of its host galaxy: very massive black holes are found in galaxies containing very many stars; smaller black holes tend to inhabit more sparsely populated galaxies. Could this be evidence that the rate of growth of a black hole is determined by the galaxy in which it resides?

Black holes with a mass of just a few times that of the Sun (say up to 100 times the solar mass) will form as a result of the supernova explosion and core-collapse of massive stars. Really massive stars (more than 100 solar masses) may go straight to the black hole stage without even exploding as a supernova. Once formed, these black holes will absorb material. Sometimes they might collide and coalesce. Inexorably, the amount of ordinary matter that disappears out of existence into black holes will increase as the universe ages. The supermassive black holes detected at the centres of galaxies are not just a few times but millions of times the mass of the Sun. Although no one knows whether they preceded the galaxies in which they sit, or whether they are the bloated descendant of an early small black hole, there has certainly been enough time since the beginning of the universe for them to have grown to their present size. As massive stars die in supernova explosions, more black holes will be formed. Those that already exist will grow ever more massive.

There is a massive black hole at the centre of our own galaxy; it is dormant but it too is absorbing material. But this does not mean that the future for our own Sun is to be gobbled up by the black hole at the centre of the Milky Way. Our solar system, located as it is in the distant suburbs of the galactic star city, is too far out (and rotating around the galactic centre with too great an angular momentum) for there ever to be a danger of us being swallowed up. The same applies for most of the stars and dust in the Milky Way.

So what, then, is the fate that lies in store for this complex universe of galaxies? The answer lies with the third observation that Edwin Hubble made in the late 1920s. He discovered that the universe is expanding and that the most distant parts are receding from us faster than those close by. Will that expansion continue for ever, in which case the universe will eventually become populated by a mixture of increasingly isolated galaxies and a multitude of black holes? Or will the expansion come to a halt, and everything start contracting again, in which case the universe may end in a single black hole at the heart of Big Crunch?

Some of the answers may lie in observing and understanding the early universe, just as the characteristics of a star at its birth will pre determine its lifetime and its lifestyle. The Hubble Space Telescope has been able to peer across space and down through time to witness the very birth of the galaxies themselves and so, perhaps, the future of the universe. It is to the pictures Hubble has taken of the uttermost margins of the observable universe – the ultimate horizon – that the next chapter of this book turns.

CHAPTER SIX

LOOK BACK INTO THE FUTURE

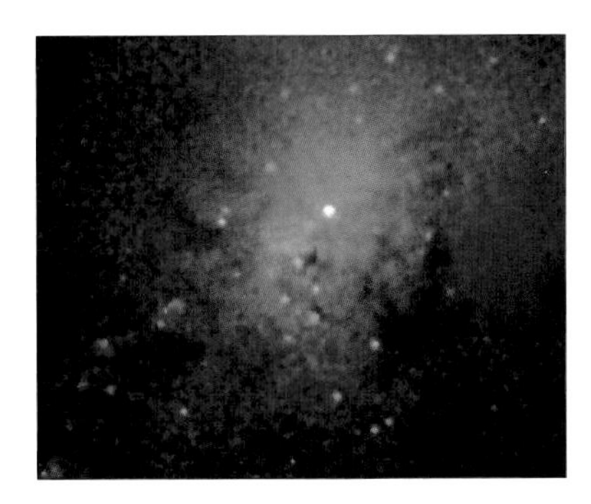

The stars are growing old. Most of the stars that will ever come into existence in the universe have already been born.

When the universe was very young, it was an active and dynamic place where stars were forming out of dust and gas at a rate ten times that seen today. The peak of star formation occurred about 3 billion years after the Big Bang that started the whole universe off. Our own Sun is a comparative latecomer, coming into existence about 5 billion years ago, by which time the stellar 'baby-boom' was over.

Opposite: For the past two decades, astronomers have wrestled with a mystery: theories of galaxy formation have long held that giant elliptical and spiral galaxies were dominant throughout the universe, yet ground-based telescopes were finding more and more signs of small blue galaxies. Research using Hubble has confirmed the growing suspicion – small blue galaxies once were the most common form of galaxy. This picture is one of more than fifty shots of the sky deliberately taken at random, which were then analysed in depth. As you can see, if you look far back enough in time, the blue galaxies are everywhere. The mystery now is, where did they all go?

Credit: R. Windhorst (Arizona State University), and NASA

Above: By contrast to the many small, oddly shaped and blue galaxies seen opposite, NGC 1275 is a giant elliptical galaxy, located at the heart of the Perseus cluster. Following on from the questions posed by the early dominance of small blue galaxies, the question is whether they ended up as the building blocks of larger, modern galaxies through a process of merger and absorption – or whether they simply faded away into obscurity.

Credit: NASA

The main image shows a deep survey made by Hubble of thousands of galaxies, ranging in age from 5 to 12 billion years old. Many of the most ancient of these galaxies are oddly shaped – astronomers came up with descriptions like 'tadpoles', 'train wrecks' and 'shards'. The image at top right shows some of these oddball galaxies tangled up together – and their very proximity boosts suggestions that the strange shapes are due to collisions and mergers. The centre right image shows a pair of ancient elliptical galaxies very similar to their present-day descendants. Bottom right is the radio galaxy used to measure the distance to the ancient cluster of galaxies.

Credit: STScI and NASA

But star formation was not the only sign of youthful activity in the universe. The peak in star-birth seems to have occurred at around the same time as the majority of quasars started to blaze out, each with the power of up to 1,000 galaxies. Incredibly, galaxies themselves appear to have formed even earlier than the peak in star formation, within 100 million years of the Big Bang. But these were not galaxies as we know them today: they were fragmentary, odd-shaped things which appear to have developed into the spiral and smooth ellipticals of the present-day universe by merger, collision and accretion.

For the first time since Edwin Hubble proved more than seventy years ago that the universe was immensely bigger than anyone had suspected, and that it consisted of vast numbers of island cities of stars, astronomers are beginning to get decent observational evidence to help them answer one of the outstanding questions in modern astronomy: how did the galaxies form in the first place?

As the preceding chapters have demonstrated, astronomers have worked out solid theories to describe most of the inhabitants of the universe. The origins of our own solar system, and why it has the structure that it does, are things that are comparatively well understood. The birth, lifestyle and death of the stars are also understood, to the extent that astronomers can apply the theories of nuclear and quantum physics devised as a result of experiments here on Earth to the processes going on inside the hearts of distant stars. The creation of the chemical elements of which we are made, the recycling of dust into stars and then back again to the interstellar medium; all these events can be modelled mathematically and on computers.

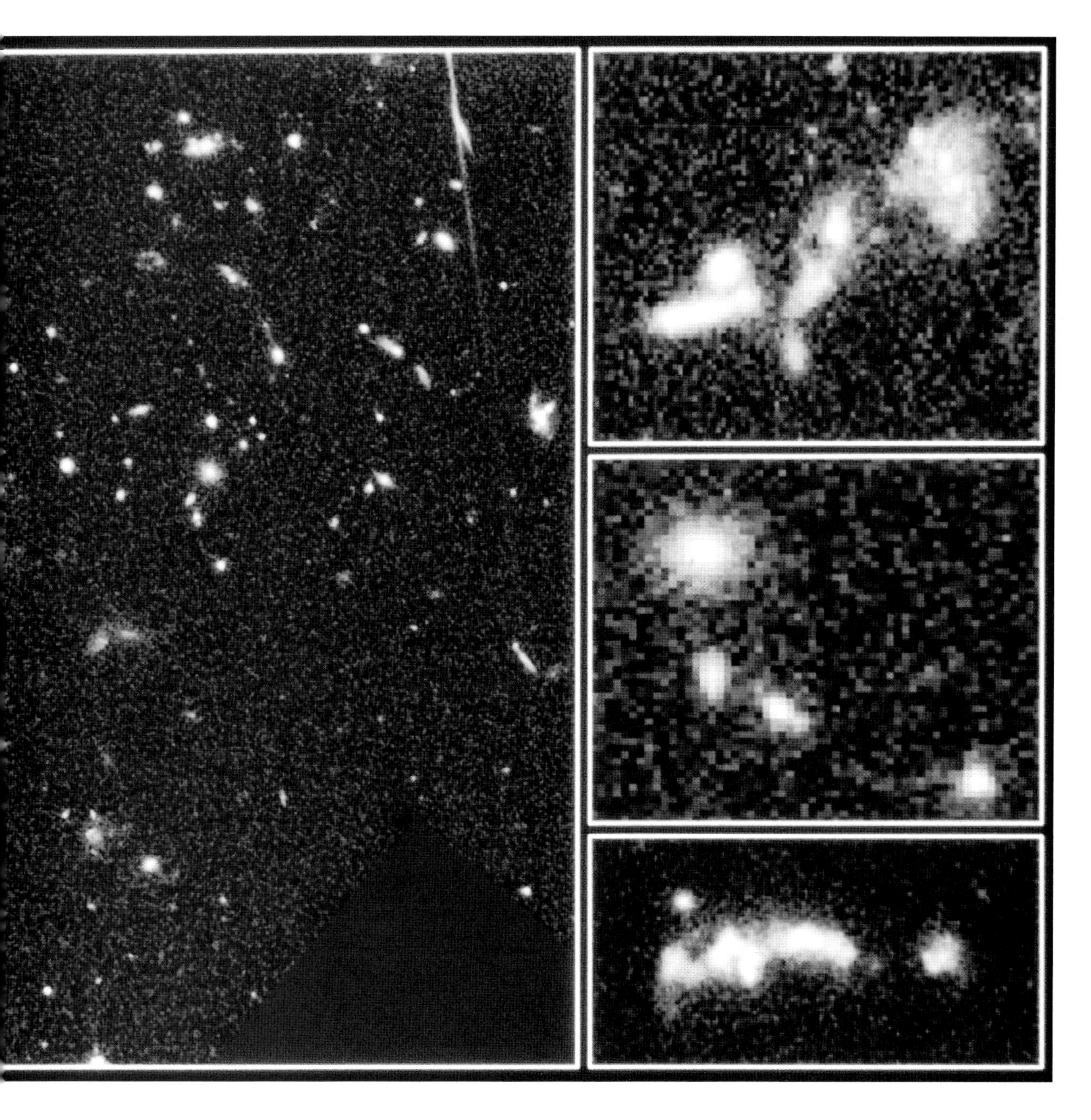

Sometimes the details are not quite right, but the overall picture is an astonishing testimony to human cleverness; an amazing amount of the universe makes sense, more perhaps than we had any right to expect.

But for the most significant objects of all in the universe – the galaxies – the situation is quite different. Whereas one can predict that the denser patches in a smallish cloud of hydrogen gas would eventually coalesce and that stars would begin to shine as the hydrogen fused into helium, no one knows why there are galaxies at all. The story of galactic astronomy is one in which observations of phenomena constantly outstripped the ability of the theoreticians to explain what was being observed.

Surprisingly, perhaps, the outline history of the

universe as a whole is better understood than the evolutionary history of the galaxies. Virtually no astronomer today seriously questions the idea that the entire universe came into being in a primordial fireball – the Big Bang – which created not just matter and energy but time and space as well. (That is why the true answer to the question 'What was there before the Big Bang?' is so psychologically unsatisfying. The answer is literally 'no thing' and also 'no time', and therefore there was no 'before' Big Bang.)

The foundation stones for this understanding of cosmology were laid in 1914 by Vesto Slipher, working at the Lowell Observatory in Arizona, who showed that eleven out of fifteen 'spiral nebulae' that he examined had their light shifted towards the red end of the spectrum. After Edwin Hubble had shown that these spiral nebulae were in fact remote galaxies in their own right, he worked for several years with Milton Humason (who had had no formal training as an astronomer and started work at the Mount Wilson Observatory as a mule train driver) on the redshift of all the galaxies they could measure. In 1929, Hubble and Humason announced perhaps the most sensational discovery of this sensational decade: the galaxies were all receding from the Earth at incredible speed and, moreover, their speed was proportional to the distance they were away from us. The most distant galaxies were travelling away from us at the fastest speeds. The signal of their motion was the fact that their light – which contains characteristic lines representing chemical elements – was all shifted from the normal part of the spectrum to the redder end. Hubble interpreted this redshift as the Doppler effect, well known when a police car's siren gets higher and higher in pitch as it approaches you, only to fall as it passes and recedes into the distance. Light too shows a Doppler effect, and the equivalent of lower pitch is redshift.

Cosmologists subsequently realised that the most natural explanation of the redshift and the speed–distance relationship was not that the galaxies were moving through space, but that *the fabric of space itself* was expanding. Imagine three dots each an inch apart painted in a line on the surface of a balloon. The dots never move relative to the surface

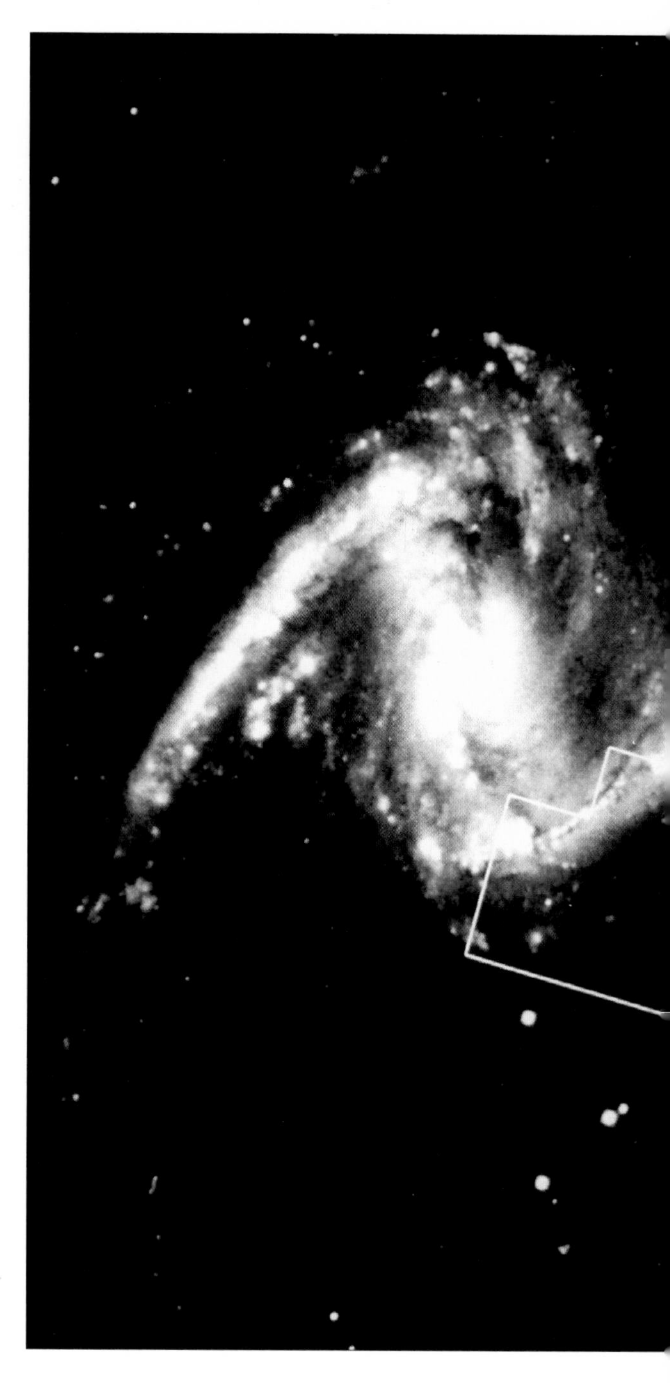

The spiral galaxy NGC 1365, which is located in the cluster of galaxies known as Fornax (and which is visible from the southern hemisphere). The mono image on the left shows the entire galaxy, as photographed via a ground-based telescope. The colour image, taken by Hubble, shows a region where young stars have been forming in the spiral arms, and which is visible as the series of bright blue dots. The Hubble team found fifty stars in the region known as Cepheid variables. These can be used to measure distances of celestial objects, from which the team calculated that the galaxy lies approximately 60 million light-years from Earth.

Credit: W. Freedman (Carnegie Observatories), the Hubble Telescope Key Project team, and NASA

This series of images helps illustrate just how hard it is to pick out the most distant (and therefore the most ancient) clusters of galaxies. The image on the left, taken by a ground-based telescope, shows a patch of sky covering the same area as the bowl of the Big Dipper constellation (the Plough). The centre image, taken by Hubble, shows a small patch of the first image, and reveals a cluster of distant galaxies – 12 billion light-years away – which were quite invisible in the first image, due to the light put out by the stars and galaxies in the foreground. Fourteen of the galaxies visible here are in the ancient cluster; the others are galaxies in the foreground. Astronomers can tell which is which by looking at the degree to which the light from any particular galaxy is red-shifted. The image at top right is an enlargement showing one of the most distant galaxies (the patch of light off-centre and to the right), which is 300 million light-years nearer to us than the incredibly bright quasar which also appears in the picture.

Credit: Duccio Macchetto (ESA/STScI), Mauro Giavalisco (STScI), and NASA

of the balloon, but as the balloon is inflated, the distance between them grows because the fabric upon which they sit is stretched. If the balloon doubles in size, the first and second dots, which start 1 inch apart, will finish 2 inches away from each other; whereas the first and the third dots, which start off 2 inches apart, will end up 4 inches apart. Thus the most distant dot appears to recede the fastest (3 inches rather than 1). Similarly the galaxies appear to move away from us because of the expansion of the universe, and their light is red-shifted because of this cosmological effect – the light gets 'stretched' in transit – rather than from the Doppler effect. The uniform expansion of the balloon in this analogy explains what at first sight appears to be the extraordinary result that the distance and redshift of a galaxy are related by a simple constant of proportionality. Multiply a galaxy's distance by a universal constant, called appropriately the Hubble Constant, and you can work out the speed at which it is receding from us. Equivalently, since redshifts are easier to measure

than absolute distances, divide the velocity by the Hubble Constant and you get the galaxy's distance.

If the galaxies are moving apart today, then it seems obvious that they must have been closer together in the past. Thus was born the idea that the universe originated in a single instant many billions of years ago in the Big Bang. Some cosmologists disputed the idea and came up with other, ingenious explanations. But these faded in 1965 when Arno Penzias and Robert Wilson built a radio telescope that heard the 'echo' of that primordial explosion. They proved that the entire universe was bathed in microwave radiation characteristic of a temperature just 2.7 degrees Celsius above absolute zero. Astronomers could find no way of interpreting the result other than as the relic of the original fireball of the Big Bang.

Other questions start to crowd in at this point. When did the Big Bang happen (i.e. how old is the universe)? What is the value of the Hubble Constant? Will the expansion of the universe carry on for ever so that each galaxy gets progressively more remote from its neighbours? If so, the universe will get emptier and colder and end in a 'Big Chill'. Or is there so much material in the universe that its gravitational pull will slow the expansion to a halt, eventually allowing the universe to subsist in a steady state? Or could it be that the same gravitational pull will reverse the process, and, like a huge piece of elastic rebounding, the entire thing will collapse back into a 'Big Crunch' and everything will be destroyed in a vast implosion which will leave only a huge black hole? Remarkably, the Hubble Space Telescope has been able to help us answer some of these questions; it has been able to act both as a window into our universe's past, discerning details about how it all began, and as a 'future-scope' to tell us how it might end.

In January 1996 perhaps the most extraordinary picture ever taken by the Hubble Space Telescope was unveiled to the excitement, amazement, and consternation of astronomers attending the regular winter meeting of the American Astronomical Society.

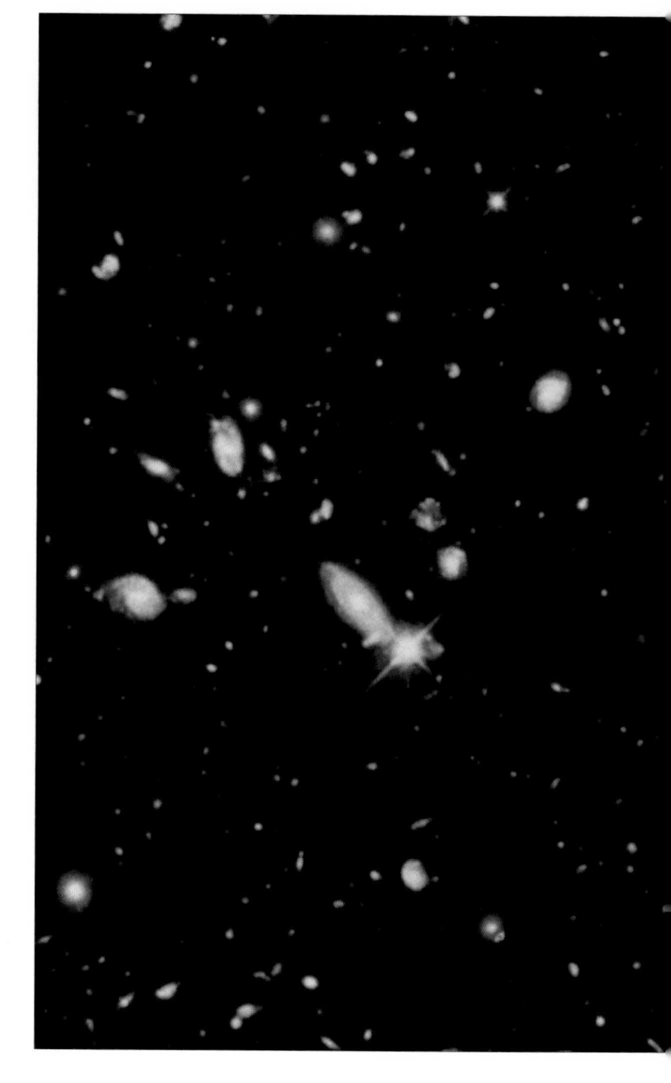

The image of the Hubble Deep Field, as it was called, revealed a bewildering array of more than 1,500 galaxies at the uttermost edge of the observable universe. Of course, there was no way to measure the distance directly; that they were very distant galaxies was recognised, thanks to Edwin Hubble, from the extreme degree to which their light was redshifted. The faint light from the most distant of these galaxies was too feeble to

show up against the background glow of the Earth's atmosphere and so was never going to be seen by ground-based telescopes, no matter how large they might be. Only the orbiting Hubble Space Telescope could return the haunting images of the distant universe to human eyes. It had seen to the outer limits of space – and also of time.

The light from the galaxies pictured in the Hubble Deep Field has taken 10 to 12 billion years to reach us; it therefore carries information about those galaxies not as they are today, but as they were 10 to 12 billion years ago, soon after the universe was created. Thus, for the first time ever,

A detailed section from the famous 'Hubble Deep Field' picture which stunned scientists when it was first released. It showed at least 1,500 galaxies at all stages of evolution. Most were so distant and faint that they had never before been seen. Their very distance indicates that they are galaxies as they were at the dawn of time – within a mere billion years of the Big Bang that began the universe and time itself. The extraordinary picture was created by assembling 342 separate images, taken over a 10-day period, on a minute scrap of sky. So small, in fact, that it covers about the same area as the diameter of an American dime seen 75 feet away. This method allowed astronomers to 'tunnel' deep into the sky to extract even the faintest images.

Credit: R. Windhorst (Arizona State University), and NASA

The early universe contained a bewildering variety of galaxy shapes. As well as the familiar spirals and ellipses of the night sky of today, the deep probing of the Hubble team found many great gatherings of stars in galaxies of irregular and peculiar shape. The images taken by the team were made available to scientists all over the world, so that they could begin detailed work on them. The Deep Field Survey may prove to contain some of the most important evidence about the past development of the universe since Edwin Hubble himself made his ground-breaking discoveries more than seventy years ago.
Credit: Robert Williams / Hubble Deep Field Team (STScI), and NASA

astronomers had pictures of truly young galaxies.

Theories about the evolution of galaxies have been difficult to formulate, and almost impossible to check. In contrast, it has been relatively easy to check theoretical predictions about the evolution of stars against observable evidence. Stars age too slowly for astronomers to track stellar evolution in 'real time', but there are so many stars in the Milky Way at different stages of development that it is possible to observe all the phases of stellar evolution by looking at different stars. In that sense, we are fortunate to be living in a spiral galaxy where stellar formation is proceeding apace; had our solar system come into existence in an elliptical galaxy where most of the stars formed early in the galaxy's history and had aged to red giant status, it would have been more difficult to check astronomers' theories about stellar evolution. But since Edwin Hubble's announcement in 1924 that the universe was constructed of galaxies, astronomers have been able to 'see' events only in comparatively nearby galaxies. This means that they have had to make their observations from galaxies which are all (more or less) the same age as our own, and therefore nothing very different is happening in them against which to compare theories.

All this changed on 15 January 1996. The image that was released to the world that day had been painstakingly constructed over ten days before and during Christmas 1995. Given the constraints on the availability of observing time available for the Hubble Space Telescope, it had been a bold move to allow one rather speculative programme to monopolise one camera for so long. The decision was a personal one taken by Dr Robert Williams, who was then the Director of the Space Telescope Science Institute in Baltimore, USA. He called together a special advisory group of astronomers, which looked at other evidence – including the more modest 'medium deep' survey and some ground-based observations – and recommended that Dr Williams should use much of his own discretionary time on the telescope for this purpose. Scientists tend to be conservative about their science; they do not make progress in leaps and bounds, no matter how much science is presented in newspapers and on television as a series of breathless 'breakthroughs'. Rather, scientific advance is a series of very small steps, each shuffling just a little bit further than the one before. There had been hints from earlier images that the early universe contained strangely shaped galaxies. So Dr Williams and his advisory group had a good idea that they would find something of interest in a deep field survey. But it was not a sure-fire thing; it could so easily have been the case that the deep field survey revealed little or nothing, either because no galaxies had started to form at that early stage in the universe's history, or that the young galaxies would be the same shape as those we can see today in the local neighbourhood.

During the ten days between 18 and 28 December 1995, the Wide Field and Planetary Camera 2 was kept trained on a narrow patch of sky in the constellation of Ursa Major, the Great Bear. ('Narrow' here means smaller than the amount of sky blotted out by a single grain of sand held at arm's length.) These observations are analogous to the core of rock which geologists extract by drilling exploratory boreholes down into the interior of the Earth. Geologists can decipher both the history and the structure of the Earth by observing the sequence of strata which have been laid down one on top of the other. The Hubble Space Telescope looked all the way down a narrow 'tube' of space, capturing images of galaxies both utterly remote and somewhat closer – galaxies ancient and modern. To help weed out excessive clutter from 'foreground' objects, the field of view was deliberately sited away from the mass of the Milky Way.

Each of the 342 separate exposures – taken in ultraviolet, blue, red, and infra-red light – took between fifteen and forty minutes. The astronomers at the Space Telescope Science Institute then combined these different exposures into one final picture. When it was complete they had the deepest picture ever taken of the cosmos. Geologists often find the remains of long-dead creatures in their core rock samples, the precursors of the plants and animals that are alive on Earth today. The fossil remains seldom look like living creatures because biological evolution has changed their forms unrecognisably over the millions of years since multi-cellular life started on Earth. Similarly, the Hubble Deep Field has yielded 'fossils' of galaxies, capturing their likeness as they were long ago, before galactic evolution set in and changed their shape totally.

The image showed that Dr Williams' gamble had paid off handsomely. The juvenile galaxies visible from so long ago included familiar ellipses and spirals, but there were also some galaxies shaped like beach balls and footballs, and others were long cigar-shaped clusters of stars. Rather than holding on to the data for the team who had made the observations to analyse, it was released to the entire astronomical community to pore over and tease out the information. The picture stunned scientists because it challenged accepted theory: before they saw the Deep Field, many astronomers had believed that at the time the images were generated, the universe would have been too young for galaxies to have formed at all.

Although the Hubble Deep Field as presented to the American Astronomical Society meeting on 15 January 1996 (and shown here) is a single image, it is a construction which represents many different individual observations, each of which was seized upon by teams of eager astronomers. Years of analysis will follow before they have squeezed every drop of meaning from the data that Hubble has sent back. In the end, the Deep Field images may well come to be seen as among the most important astronomical observations made since Edwin Hubble's original discoveries about galaxies in the 1920s. It certainly vindicated the original scientific purpose of the building and launching of the Hubble Space Telescope: to measure the size and age of the universe, and to throw light on the emergence and evolution of the galaxies.

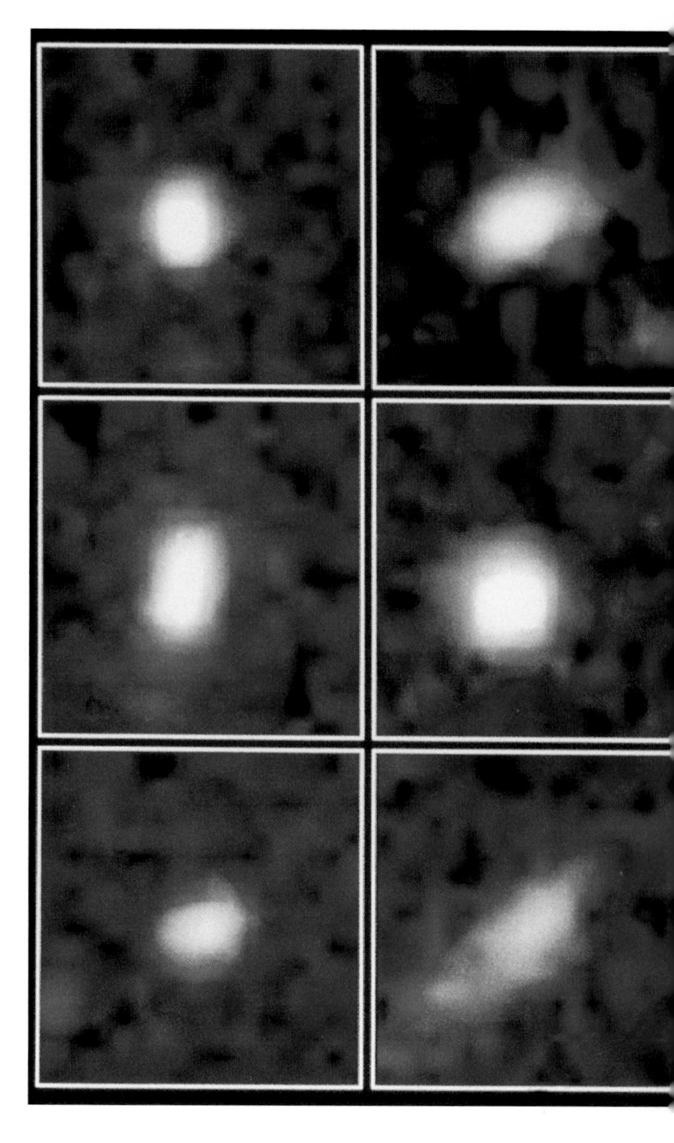

A year after the images had been taken, one team of astronomers announced that they had

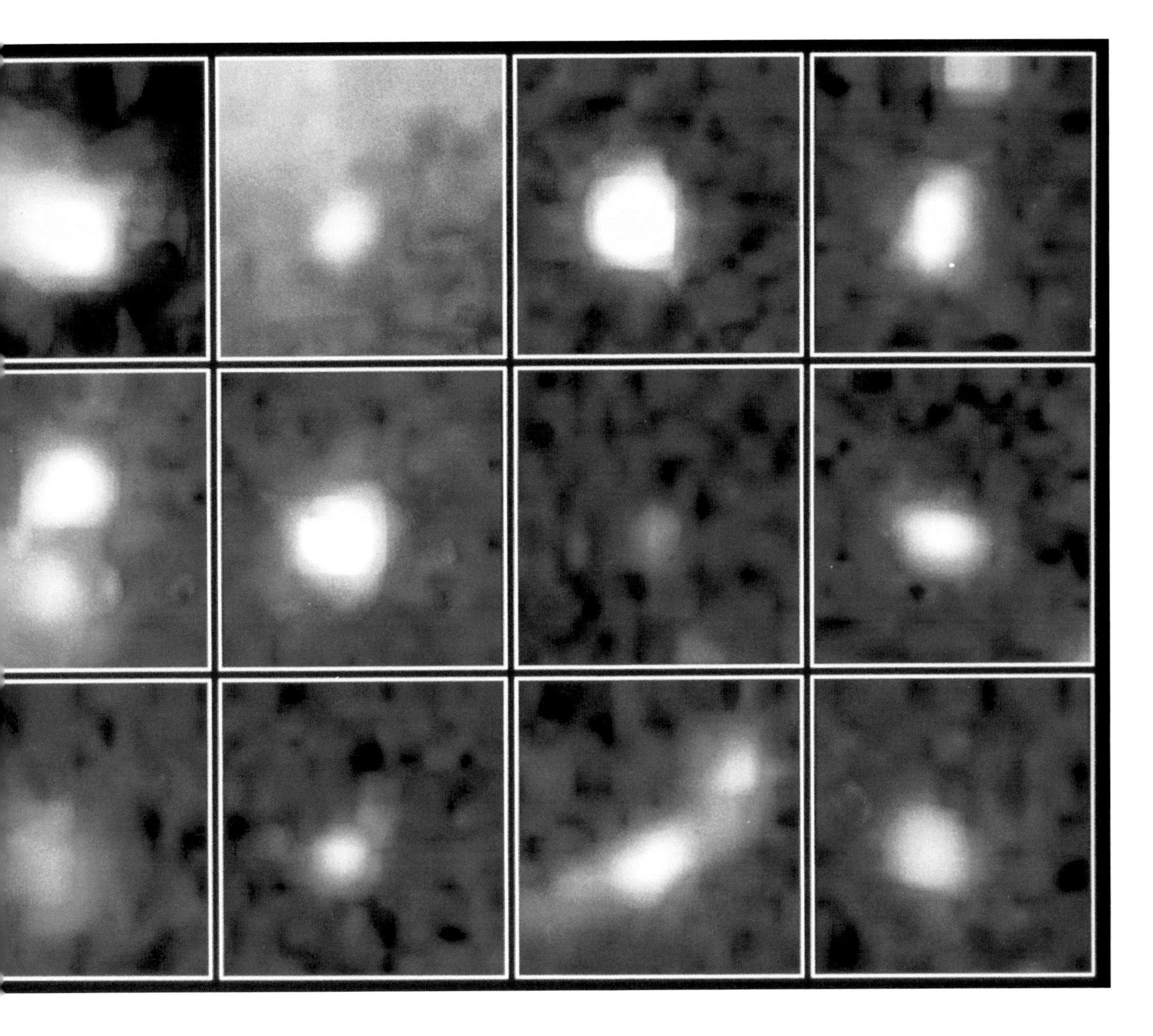

analysed the colour of the light from some of the most distant galaxies. It contained evidence showing the rapid burst of star formation that occurred in the early universe – the stellar 'baby boom'. The researchers used an unusual technique to estimate the distance to the galaxies that they were examining, comparing images taken in visible light with those taken in ultraviolet. Many visible galaxies are not present in the images taken at ultraviolet wavelengths because of the vast clouds of intergalactic hydrogen gas through which light from the distant galaxies must pass to reach us. The

Theories that today's galaxies were formed by collisions and mergers between smaller galactic 'building blocks' received a boost when Hubble found eighteen gigantic clusters of stars packed into a relatively small area of space – and certainly close enough together to have merged into one or more much larger true galaxies. These 'proto-galaxies' are predominantly blue in colour, suggesting the presence of many hot, massive stars in the process of formation. Their distance indicates that they date from around 11 billion years ago, about the time that many theoreticians believe true galaxies started to form. Each is about 2,000 light-years across, compared with the 100,000 light-year diameter of our own Milky Way galaxy.

Credit: R. Windhorst (Arizona State University), and NASA

The galaxies contained in this Deep Field Survey image have been pictured at a time when the universe was only about 16 per cent of its present age. This demonstrates one of Hubble's most valuable characteristics – by focusing on the same types of objects in different parts of the sky, or at different depths within the same patch of sky, the Space Telescope can show different stages in evolution of the object in question – hence the description of the telescope as a 'time machine'. Its ability to look far back into the past – to almost the beginning of time itself – means that it may end up providing concrete answers to the great questions that have puzzled man for millennia.

Credit: Rogier Windhorst and Sam Pascarelle (Arizona State University), and NASA

hydrogen selectively absorbs ultraviolet radiation, while letting visible light through. In this way it is possible to distinguish distant from 'nearby' objects. However, the expansion of the universe means that much of the radiation that started out from these galaxies as ultraviolet was shifted in transit to visible wavelengths by the time it entered the Hubble Space Telescope. By analysing the spectrum of this light, the astronomers could correct for the expansion of the universe and calculate how much ultraviolet radiation had set out on its long journey to Earth. Ultraviolet radiation is a hallmark of young, new-born stars, and so the total amount of ultra-violet light indicates the degree to which hot, young and massive stars are present in the galaxies being observed.

Starbirth, in its turn, can be an indicator of intergalactic collisions, as was described in the previous chapter. Could it be that the formation of galaxies as astronomers know and see them today was the result of massive collisions in the very early universe? The results from the Deep Field are suggesting that this was indeed the way in which some galaxies evolved. Many astronomers now believe that only small galaxies existed in the very early universe, and that these represent the building blocks out of which were formed the large, mature galaxies we can see today. The immense power of gravity, acting slowly but inexorably over billions of years, drew many of these small 'proto-galaxies' together so that they coalesced into large star conurbations. The most distant objects visible in the Deep Field are also among the most strangely shaped; as the most distant, they would also be the ones furthest back in time. A quick glance at a textbook on the evolution of life on Earth shows that the ancestors of the animals we see on the planet now were strangely shaped creatures indeed. Evolution on a galactic scale means that these objects pictured in the Deep Field are the ancestors of the galaxies and, like our own ancestors, are both weird and long since extinct.

This evolutionary or merger hypothesis remains only a theory, and one contested by some astronomers. But a second look at distant galaxies, albeit ones not quite so far away as the Hubble Deep Field, has boosted the merger theory. In September 1996 American astronomers announced that, with the help of Hubble, they had caught eighteen 'proto-galaxies' in the act of merging into a couple of large galaxies about 11 billion years ago, in a small region of sky to the north of the constellation of Hercules near where it joins Draco, the Dragon.

The lengths to which the scientists had to go to detect these galactic building blocks shows both how difficult galactic astronomy is, and how astronomers will deploy all the means at their disposal to pin down elusive fragments of data that will help them decipher the puzzle of galactic origins. As well as the Hubble Space Telescope, they used the Multi-Mirror Telescope on Mount Hopkins south of Tucson in Arizona, and two telescopes on Mauna Kea, the NASA Infra-Red Facility and the W.M. Keck Telescope. (The ground-based observations were critical in proving that the clusters of stars being observed were all at about the same distance. They were used to analyse the spectrum of the light to show that it had all been redshifted by the expansion of the universe to the same degree, and therefore that the radiation had all travelled the same distance.)

Each of the 'proto-galaxy' building blocks appears to consist of about 1 billion stars, with all eighteen packed into a space about 2 million light-years across – about the distance separating the Milky Way from Andromeda, the nearest major galaxy. The objects appear to measure only about 2,000 light-years across (compared to the 100,000 light-year diameter of the Milky Way) and to be crowded with young blue stars and clouds of glowing gases. Four of them appear to have double structures at the centre, suggesting that they themselves are the product of recent mergers.

In January 1998, two years after the images were first released, Steve Zepf from the University of California at Berkeley announced that he had found evidence also supporting the galactic merger hypothesis. Analysing the Deep Field survey, Professor Zepf discovered that elliptical galaxies – which are such dominant members of the modern universe – were pretty much absent from the very early universe as seen in the Hubble Deep Field. Elliptical galaxies are characterised by predominantly old and cool stars, and so tend to shine strongly in the infra-red region of the spectrum. This effect is enhanced by the way in which the radiation from very distant galaxies is anyway shifted to the red end of the spectrum as the universe stretches. So Professor Zepf compared the Hubble Deep Field image with data from infra-red telescopes and found that galaxies with colour signatures characteristic of modern ellipticals were almost completely absent from the population of galaxies in the early universe as imaged by the Hubble Deep Field. There were far fewer very red galaxies than would have been expected from the size of the modern population of elliptical galaxies. The obvious conclusion is that the ellipticals had not yet had time to form when light from the faintest and furthest objects in the Hubble Deep Field started its journey to Earth. Ellipticals must have formed later by merger and accretion of the proto-galaxies.

However, the idea that galaxies evolved by

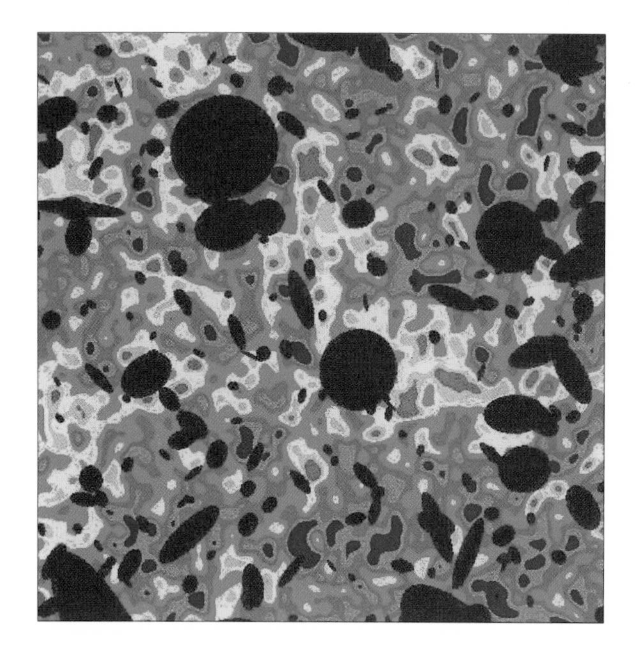

Above: This extraordinary image is the result of a fine piece of astronomical lateral thinking. Whereas most research teams have been studying the objects found in the Deep Field images, Dr Michael Vogeley of Princeton University decided to ignore the objects, and to look at the blank spaces between the objects in an effort to discover variations in background light which would reveal whether Hubble was missing many more and even fainter galaxies. The variations were tiny – showing that Hubble is seeing just about everything there is to see.

Credit: Michael S. Vogeley (Princeton University Observatory), and NASA

merger is still disputed by some astronomers. They argue that the great age of the stars in elliptical galaxies, and their similarity in colour and brightness, suggest that they were all formed very early on and over a short period of time. Some astronomers have suggested the collapse of a huge cloud of gas as a possible mechanism, so that a galaxy-sized population of stars forms more or less at the same time, rather than a galaxy being created by the gradual merger of smaller clumps of stars. The jury is still out, for there are possible complications (dust clouds etc.) that could interfere with the interpretation of Professor Zepf's result. But recent research indicates that, thanks to the discovery of these galactic fossils, at long last astronomers are on track to developing a theory of evolution of galaxies that will answer some of the most fundamental questions about these most immense structures of the universe.

The Deep Field Survey has also led astronomers to quadruple their estimate of how many galaxies there might be in the universe. Although no one can know for sure, some astronomers now estimate there may be about 200 billion galaxies in the universe, rather than the 50 billion that was once thought to be the case. For comparison, there are about 100 billion stars in the Milky Way, so there are two galaxies in the universe for every star in our own galaxy. This would mean that there could be as many as 20,000,000,000, 000,000, 000,000 stars – 20,000 billion billion – in the universe. However, the question immediately arises: could there be even more? Since the Deep Field survey has sprung such unexpected observations on astronomers, could there be yet more galaxies so far away (and therefore so faint) as to be just out of sight even of the Hubble Space Telescope, but which a more far-sighted next-generation telescope might be able to pick up? Might the estimate for the population of the universe have to be revised yet again?

Two years after the release of the original Deep Field picture, Dr Michael Vogeley of the Princeton Observatory in the USA answered the question: no. He concluded that the Hubble Space Telescope has found most of the visible light in the universe. While armies of excited astronomers had been studying the peculiar images of the galaxies in the Deep Field picture, Dr Vogeley had turned his attention to the spaces in between them, where nothing was to be seen. He analysed these apparently blank patches for the slightest variations in brightness that might be the signature of even more galaxies; too faint to be seen as such, but just

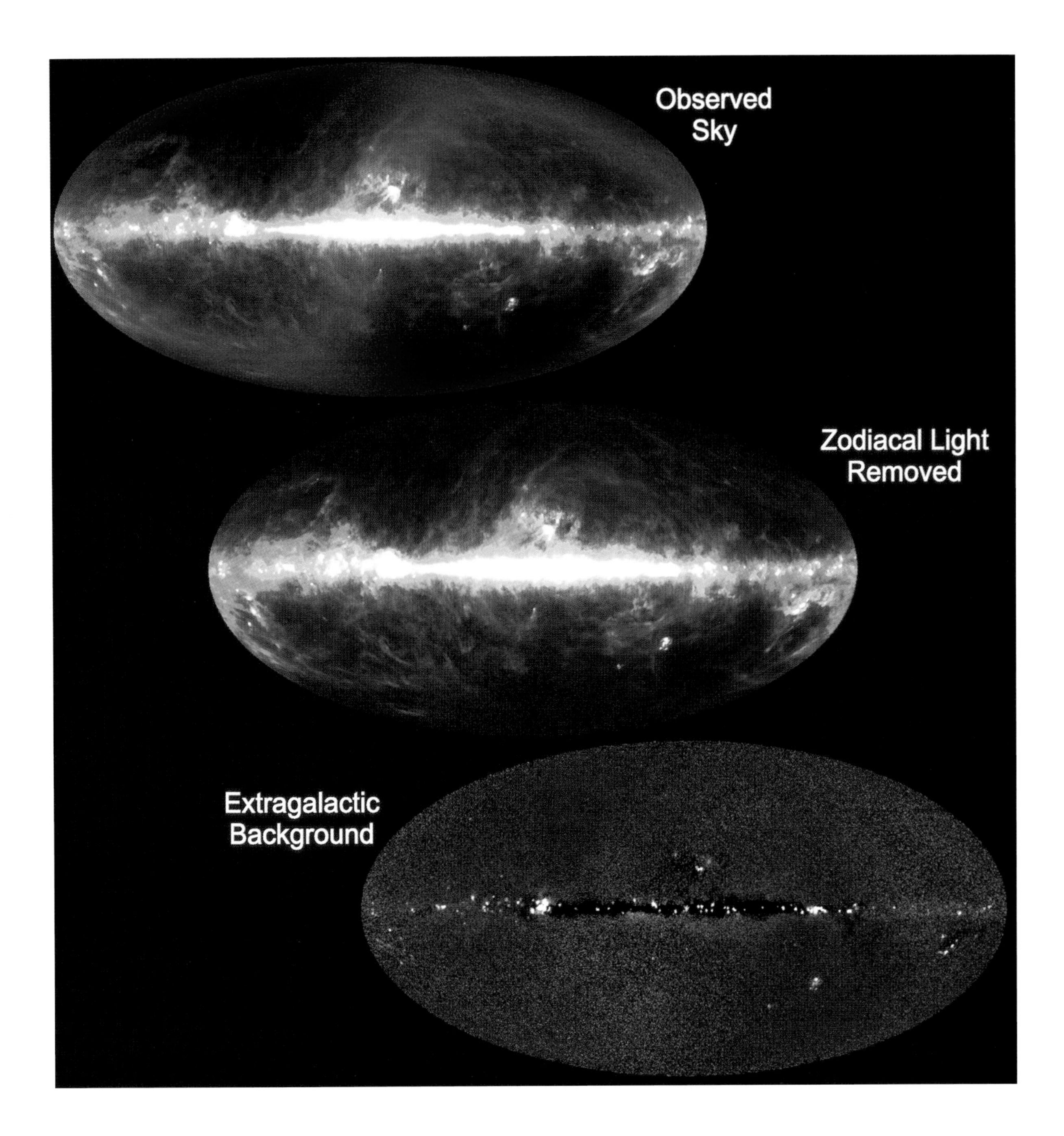

Just as the picture opposite is drawn from the search for all the visible light in the universe, so these images were produced by astronomers searching for all the infra-red radiation in the universe. They found far more background heat radiation than could be accounted for by all the stars visible in the sky. This 'fossil' radiation suggests either than many stars are hidden behind clouds of dust and gas, or that the early universe saw a burst of star birth and death, whose only memory now is the telltale infra-red trace left by countless billions of extinct stars.

Credit: M. Hauser (STScI), and NASA

luminous enough to disturb the brightness of the empty parts of the sky. He found that the brightness varied by only about one part in a thousand, not enough to suggest the existence of as yet unseen galaxies. 'The extraordinary smoothness of the background sky suggests that most of the visible light in the universe hails from galaxies that Hubble can detect,' he said.

But, as so often in science, that is not the end of the story. Dr Vogeley's analysis accounts for the visible light, but it is still possible that many galaxies are hidden from view by clouds of dust and so would be seen only in infra-red light. It is also possible that some galaxies might be so distant that while their light started out in the visible part of the spectrum when it started its journey to Earth, the expansion of the universe over the past 10 billion years or so may have 'stretched' it so far that it became red-shifted into the infra-red part of the spectrum.

Another satellite observatory, already in orbit, has picked up evidence of an infra-red background glow right across the universe, coming from interstellar and intergalactic dust heated by radiation from all the stars that have ever been created. And it found more infra-red radiation than can be accounted for by the galaxies Hubble has seen in the Deep Field survey. The Space Telescope Science Institute likened the discovery of this 'fossil radiation' to switching out all the lights in a bedroom only to find that the walls, ceiling and floor are glowing with eerie luminescence. The results come from the Cosmic Background Explorer Satellite (COBE) which was launched in 1989. COBE originally was set to measure the echo of the Big Bang itself – a universal background of microwaves (high energy radio waves) which permeates all of space – and discovered 'ripples' in this cosmic background radiation which reflected early irregularities in the distribution of matter and radiation coming out from the Big Bang. Eventually, astronomers believe, these irregularities would develop into galaxies and clusters of galaxies.

COBE's discovery of so much infra-red background radiation suggests that many stars may have been missed in visible light surveys, such as the Hubble Deep Field. One possible explanation is that vast blankets of dust obscure the missing stars and galaxies from our view. Another is that the missing stars were born very early on and soon cooled, so that although they continued to give out heat they were no longer hot enough to glow visibly. The Next Generation Telescope which NASA is proposing as a successor to the Hubble Space Telescope would be fitted with infra-red detectors to pick up precisely this sort of effect. Plans are also afoot for other infra-red orbiting telescopes which would further analyse this radiation to finalise the census of the stars and galaxies.

Missing light is, however, as nothing compared to the implications of some earlier observations with the Hubble Space Telescope. These observations appeared to show that half the lifetime of the universe was missing. Their implication was that the universe was much younger than anyone had thought – perhaps as little as 'only' 8 billion years old. The problem is that astronomers are reasonably confident that they can work out fairly precisely the age of some ancient stars – and the answer comes out at 13 to 14 billion years old. Thus Hubble measurements would make the universe younger than some of the stars that are in it: a logical impossibility. As noted in Chapter One, when the observations and calculations were published in 1994, one astronomer remarked publicly that the Hubble observations meant that 'the universe is in crisis'.

The 'crisis' happened as a result of astronomers following Edwin Hubble's footsteps. As we described in Chapter One, Hubble first measured the distance to Andromeda by measuring particular

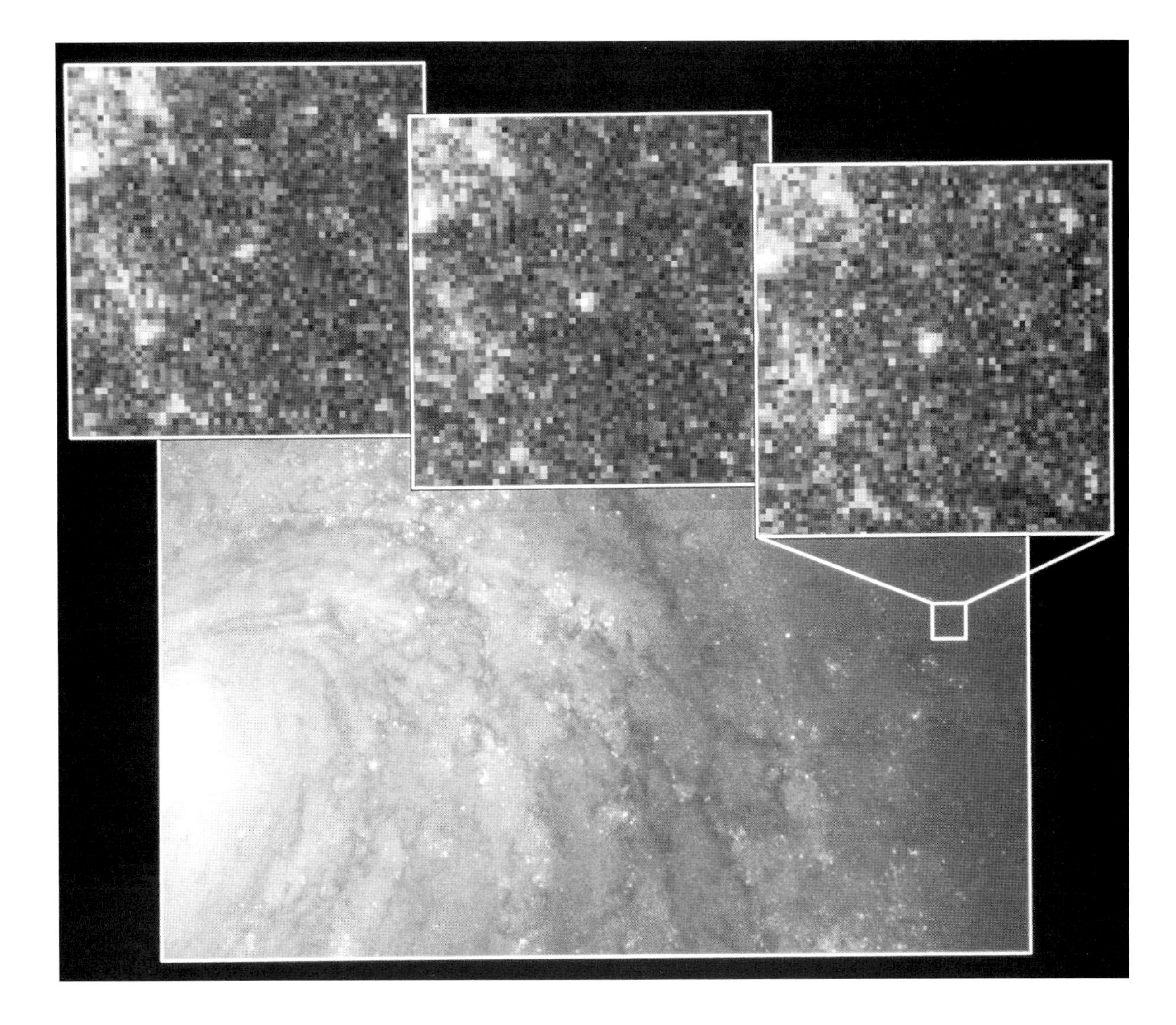

variable stars, known as Cepheid variables. These stars are useful because the rate at which they vary indicates how intrinsically bright they are. By comparing their brightness as observed here on Earth with their intrinsic brightness as calculated from the rate of variation, astronomers can work out how far away they are. (This is just an extension of the fact that a 100-watt light bulb placed 200 metres away will appear four times less bright than a 100-watt bulb placed 100 metres away, even though they are shining with the same intrinsic brightness because both are 100-watt bulbs.)

The space telescope named after the great

This series of pictures provides a close-up of the variations in brightness of the class of star known as Cepheid that has allowed astronomers to calculate precisely the distance of galaxies. The main image shows one of the spiral arms of the galaxy M100, where a good deal of starbirth is taking place. The square box shows an area that contains a Cepheid variable star, spotted by astronomers using Hubble. The three inset images show how the Cepheid varies regularly in brightness over a period of 22 days. Only Hubble has the sensitivity to pick out these 'cosmic milepost' stars. Finding more of them in the most distant galaxies will help provide secondary confirmation of their age as established by their red-shifts.

Credit: Dr. Wendy L. Freedman (Observatories of the Carnegie Institution of Washington), and NASA

astronomer has been able to pick out Cepheid variable stars in M100, a member of the Virgo cluster of galaxies. Using these stellar mileposts, a team led by Dr Wendy Freedman of the Observatories of the Carnegie Institute in Washington worked out that the M100 galaxy had to be 56 million light-years away (although there was an uncertainty either side of 6 million light-years, so M100 could be as close as 50 million or as distant as 62 million light-years). By measuring the redshift in the light coming from the galaxy, which gives the speed at which is it receding from us – and with the independent measure of the *distance* provided by the local Cepheid variables – they were then able to work out the Hubble Constant, the rate at which galaxies are moving away from us, and therefore the rate at which the universe is expanding.

By measuring the Hubble Constant, astronomers can also work out the age of the universe. It is a simple calculation, in theory: the age of the universe is given by the inverse of the Hubble Constant.

Perhaps the best way of thinking about it is the following – although this explanation is not at all rigorous, it might help to show why the Hubble Constant and the age of the universe are so closely connected. Recall what was said earlier in this chapter: divide the velocity with which an object is moving by the Hubble Constant and you get its distance from Earth. In theoretical terms, the most distant object that we could possibly observe – and

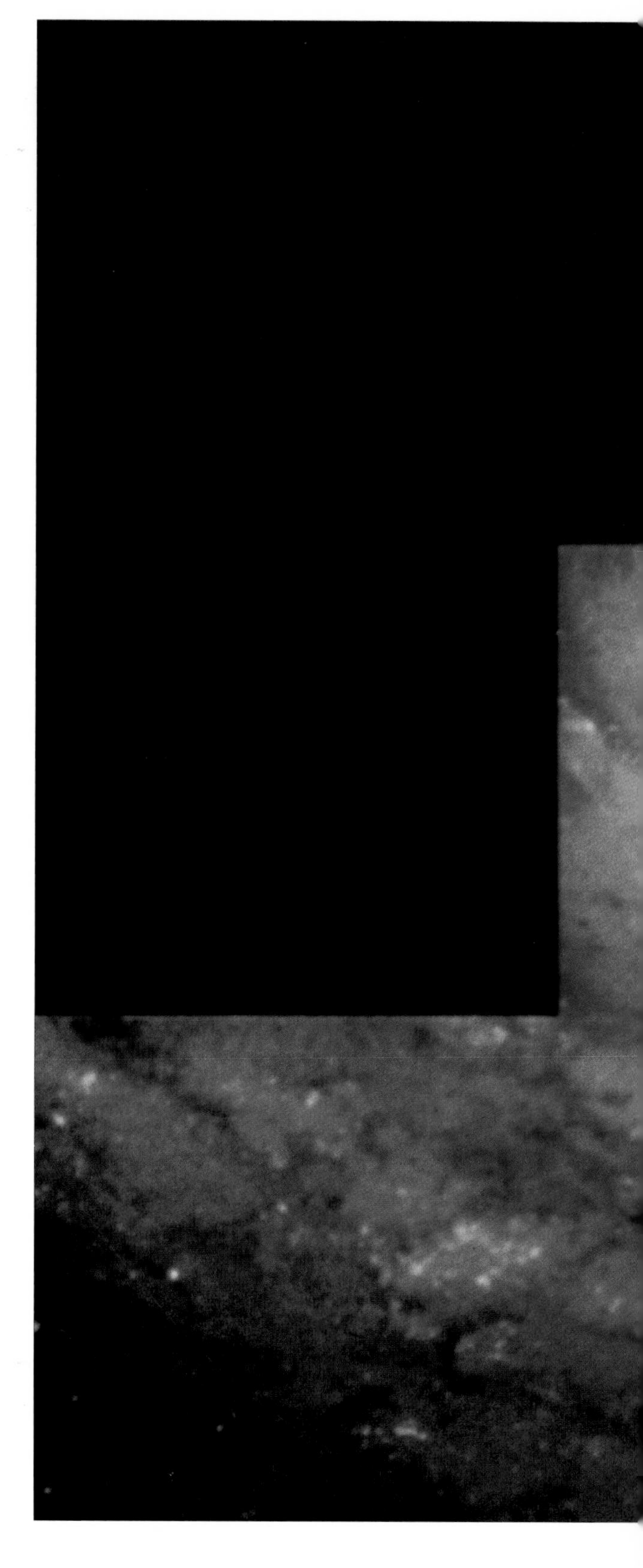

The magnificent M100 spiral galaxy, which appears full-on when viewed from the near-Earth neighbourhood. When pictured via ground-based telescopes, the distortion and turbulence caused by peering through the oceans of gas cloaking the planet prevents astronomers from seeing anything beyond a blur of light. Yet, as this image shows, Hubble can resolve individual stars within M100, allowing astronomers to study each one in detail.

Credit: J. Trauger (JPL), and NASA

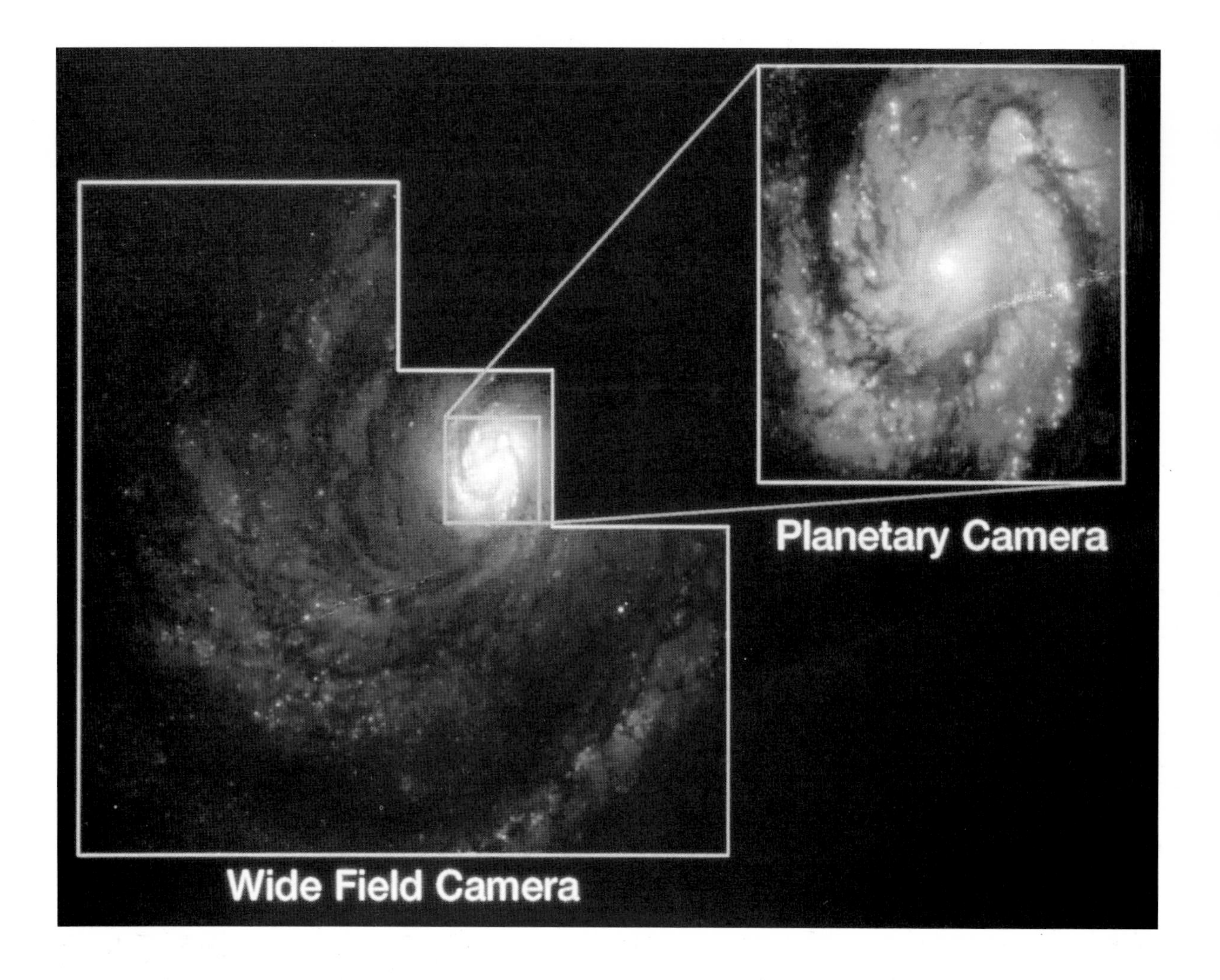

Images like this show how the Hubble Space Telescope has fulfilled its primary mission – which was to measure the size and age of the universe, to test theories about its origins in the Big Bang, and to work out how today's galaxies came to evolve. To do this, scientists literally needed to have a better picture of the universe. Using the instruments carried aboard Hubble, it was possible to establish the distance of this galaxy, M100, as 56 million light-years (plus or minus 6 million light years). By comparing this with its red-shift value, astronomers can work out the value of the Hubble Constant, i.e. the rate at which the universe is expanding. Pinpointing the age of observed galaxies will thus allow astronomers to answer the most fundamental questions about their evolution – and therefore the future. What will happen to the universe? Will it keep expanding forever, with galaxies drifting further and further apart? Or will the effects of gravitational attraction ultimately force the expansion to halt, before beginning a massive contraction that will end the universe in a Big Crunch? *Credit: (STScI), and NASA*

therefore one that would be right on the boundary of the observable universe – would be one that had been travelling with the speed of light since the universe began. The distance of the boundary from us is easy to calculate: it would be the speed of light times the age of the universe. But that distance is also the speed of light divided by the Hubble Constant. Hence the age of the universe is simply the inverse of the Hubble Constant. (The calculation is, of course, not that simple. Astronomers have to correct for the extent to which the gravitational pull of all the material in the universe may have been slowing its expansion, so the value of the Hubble Constant measured today may not be identical to that in the early universe.)

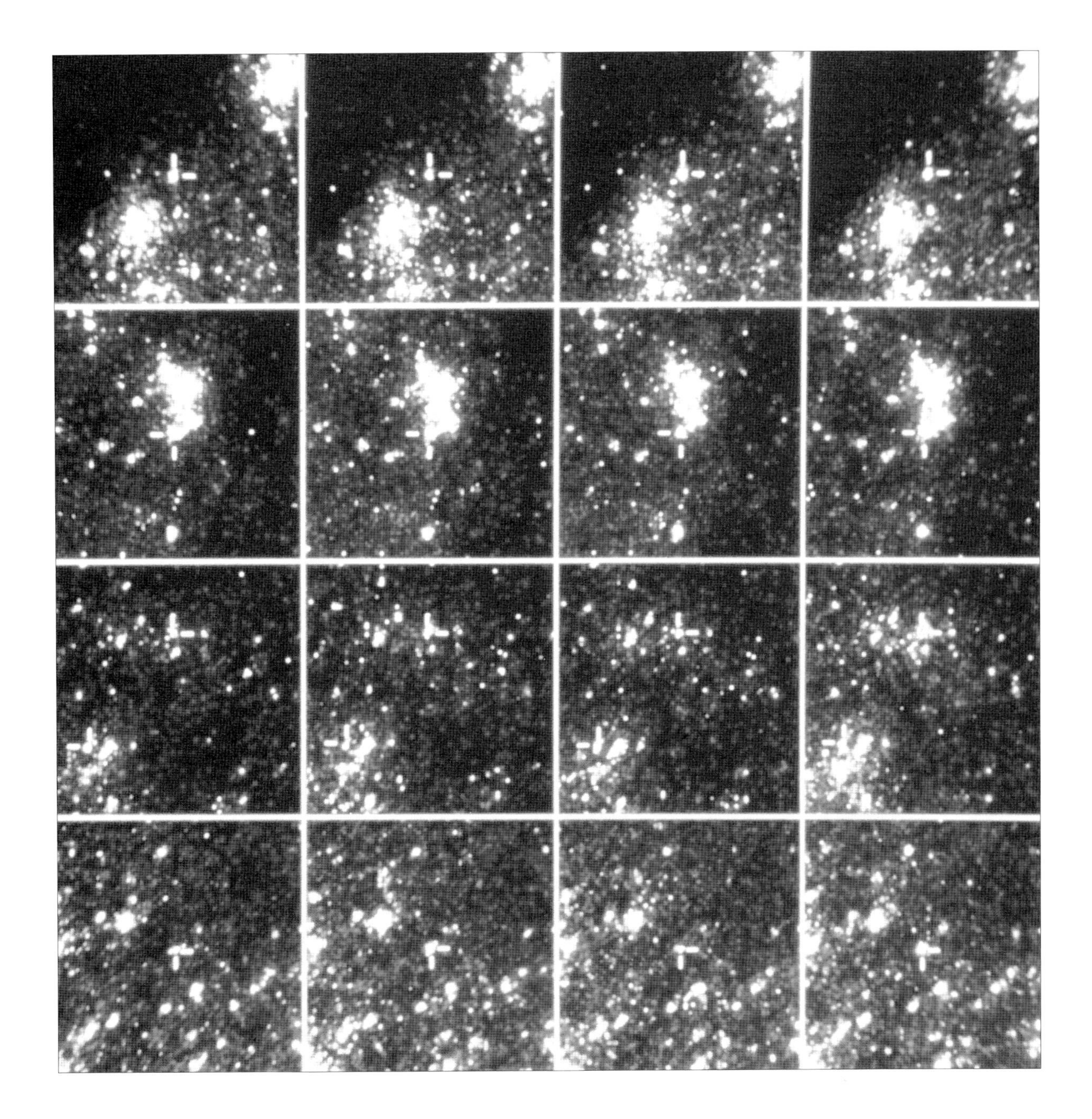

This powerful property of the Hubble Constant meant that, to measure the age of the universe, Dr Freedman and her colleagues did not need to try to measure the most distant galaxies at the limits of observational capacity. Instead, they were able to get an estimate of the universe's age by observing galaxies 'only' 50 million light-years or so distant. From this, they obtained a value for the Hubble Constant

Cepheid variables in the galaxy M81. These stars have proven extremely useful in helping to judge the distance of objects like galaxies. Because we know that the intrinsic luminosity of a Cepheid is directly related to its rate of variation, by measuring the apparent luminosity of any particular Cepheid and comparing that with its known intrinsic luminosity, we can determine how far away the Cepheid actually is – and therefore the distance of the galaxy in which the Cepheid is located.

Credit: NASA

The gravitational influence of one galaxy acts as a lens to create a magnified picture of a far more distant galaxy located behind it. The older galaxy – 13 billion light-years away – is the orange arc (lower centre right). The close-up (top-right) shows bright areas of intense starbirth. The lower right image is a computer-corrected version, removing the lens's distortion.
Credit: Marijn Franx (University of Groningen, The Netherlands), Garth Illingworth (University of California, Santa Cruz), and NASA

and worked out that the universe's maximum possible age is 12 billion years; and that, if all the matter in the universe has actually been slowing the rate of expansion down significantly, then the universe might actually be as little as 8 billion years old.

This causes problems. Astronomers believe that they understand well the life-cycles of the stars, and

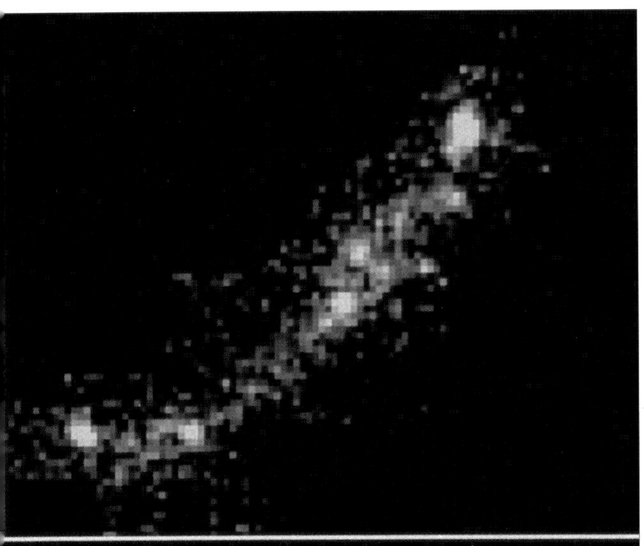

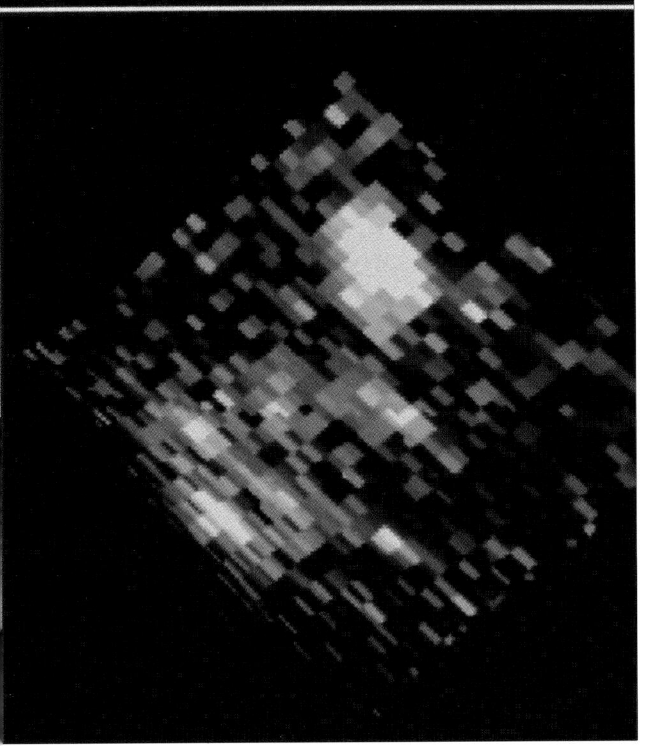

that their conclusion that some stars are *older* than 12 billion years is robust. They also find it difficult to see how galaxies could have formed and evolved to their present state quickly enough in a universe that is less than 12 billion years old (let alone one that is just 8 billion years old). The pieces of the jigsaw do not fit. So is it possible that Dr Freedman's measurement of the distance of M100 is wrong? Some scientists have speculated that the calculations could have been thrown out by unexpectedly large amounts of gas and dust between M100 and Earth, which might make the Cepheids in M100 seem fainter (and therefore further away) than they really were. Others point out that the Virgo cluster is a huge collection of around 2,000 galaxies, and suggest that the enormous gravitational attraction of this collection may be altering its speed away from us.

Astronomers need to find another way to measure the expansion of the universe so that they can check the M100 results for the Hubble Constant. In 1996 another team of astronomers, led by Allan Sandage, also of the Observatories of the Carnegie Institute, announced that using another 'standard candle' measurement they had obtained a much lower figure for the rate of universal expansion, which would give an estimated lifetime of the universe of between 11 billion and 14 billion years. This is much closer to the theoretically acceptable value. (Throughout this book, it has been assumed that the universe is something like 14 to 15 billion years old and this has been used in converting the redshifts in the light from distant galaxies into light-years. It is really only redshifts that astronomers actually *measure*; the distance in light-years is a calculation that involves an estimate of the value of the Hubble Constant or, equivalently, the age of the universe.)

Professor Sandage and his colleagues looked at a particular kind of stellar explosion known as a Type 1a supernova. This can flare up to shine brighter than an entire galaxy, but only for a period of about a month or so. Although it is called a supernova, it is not the final explosion of a dying star as described in Chapter Four but a more intriguing event still. From studying such events nearby, astronomers now believe that a Type 1a supernova results from the interaction between two stars orbiting each other in a binary system. One star is a white dwarf and the other a red giant. Very occasionally in such systems, the white dwarf's gravitational attraction pulls too much material off the red giant, triggering an explosion which tears the dying star apart in a final outpouring of light and radiation. The recoil from the explosion sends the red giant off as a 'runaway' star, several of which have been observed in our own Milky Way galaxy. The crucial point about Type 1a supernovae is that their intrinsic brightness can be measured by tracking how quickly each one fades from view. As with Cepheid variables, one can then check their intrinsic brightness against their brightness as observed from here to work out how far away they are. Type 1a supernovae are also so bright that the Hubble Space Telescope can see their explosions taking place up to 7 billion light-years away: halfway across the known universe, and more than 100 times as distant as M100.

Because Type 1a supernovae can be measured so far away, they can be used not only to check the distance scale of the universe but also to see if there is any sign of a slowdown in the rate at which the universe is expanding. In January 1998 two international teams of astronomers announced that they had analysed more than forty such supernovae and found little evidence of a slowing of the universal expansion. Preliminary results from one group, led by Peter Garnavich of the Harvard Smithsonian Centre for Astrophysics, mean that the universe could indeed be 15 billion years old. If their observations are confirmed, then the 'crisis' in the universe is over.

The supernovae may also give away the secrets of the future. Another group of scientists, led by Dr Saul Perlmutter of the Lawrence Berkeley National Laboratory in California, decided first to make sure that the characteristics of the most distant (and therefore the oldest) Type 1a supernovae were the same as the youngest (those visible in nearby galaxies). After all, they reasoned, the distant supernovae exploded when the universe was young and so there was always the possibility that their properties differed in some subtle way from supernovae exploding in the present day. Their observations confirmed that the supernovae's characteristics were similar. And that fact opened up a whole realm of scientific possibilities.

Several times in this book it has been said that the Hubble Space Telescope not only looks across space to the most distant parts of the universe, but that it also looks across time. Usually, that has meant that it looks backwards to the period when the universe was very young. But the results by the two groups led by Dr Garnavich and by Dr Perlmutter also show that the Hubble Space Telescope can look forward in time. Having established that the distant supernovae are similar to the ones nearby, as Dr Perlmutter put it: 'Distant supernovae provide natural mile-markers which can be used to measure trends in the cosmic expansion. All the indications from our observations of supernovae spanning a large range of distances are that we live in a universe that will expand for ever. Apparently there isn't enough mass in the universe for its gravity to slow the expansion – which started with the Big Bang – to a halt.'

There is a strange and humbling irony in the discovery that the future of the universe can be discerned from the death-rattle of distant stars. But it appears now to be clear from these star-destroying explosions in deepest space that we live in an 'open' universe. There will be no reversal back to a 'Big Crunch' where everything will be squeezed out of existence and the cosmos ends as a gigantic black hole. Nor does it appear as if the universe has a 'critical' density that would slow the expansion virtually to a halt without triggering a collapse. Instead, the separation between the galaxies will continue to grow and the universe will become an ever colder and ever more sparsely populated place, expanding for all time.

It is a grim picture; one in which the great star cities become increasingly isolated, marooned for ever, with their populations of stars burning out and dying away, one by one. But this may not be the end of the story. As the last chapter shows, there may be more to the universe than meets the eye.

CHAPTER SEVEN

THE LAST HORIZON

This final chapter may serve as something of an antidote to what has gone before. In previous chapters we have paid tribute to the skill and intuitive brilliance of the world's scientists, who have deduced where they could, guessed where deduction was impossible and, very often, been proved right by subsequent observations. So it is salutary to recall that human knowledge of our universe is still deficient in some important areas.

Opposite: A mosaic of different photos taken by Hubble of the area near the Coma Cluster. The brightest object in the combined picture is the elliptical galaxy NGC 4881. The main purpose of the observation was to hunt for globular star clusters surrounding NGC 4881. These are dense gatherings of ancient stars which ended up being drawn into galaxies, and which can be used to judge the distance of the host galaxy. The Coma Cluster is a great ball-shaped gathering of galaxies 300 million light-years from Earth. The spiral galaxy visible on the right of the image belongs to the Coma Cluster. Most of the other galaxies visible in the picture lie far in the background.

Credits: W. Baum (University of Washington), Hubble Space Telescope WFPC team (STScI), and NASA

Above: This striking Hubble image of the core of the peculiar galaxy NGC 7252 reveals a mini-spiral disk of gas and stars – plus about 40 globular star clusters, some of which are visible as the bright spots of light. The entire picture comprises an area 46,000 light years across; that is, were one able to travel at the speed of light (i.e. 186,000 miles per second), it would take 46,000 years to cross from one side of the picture to the other.

Credit: B. Whitmore (STScI), and NASA

Ask, for example, whether we are alone in the universe, and scientists can only shrug and quote probabilities at you. Ask if we can ever escape the appalling loneliness imposed by the sheer distances involved in crossing space, and they will talk in terms that sound like the most lurid science fiction (as we shall see later on).

And, perhaps most disturbing of all, astronomers admit that they cannot find most of the universe. More than sixty years ago the Dutch-American astronomer Jan Oort discovered something odd about the movement of some stars in the Milky Way. He looked at how they moved in directions at right angles to the galaxy's disk. To his surprise, he found that they were acting under the pull of gravity greater than could be accounted for by all the stars he could see. In fact, there had to be twice as much material around as he could see in the form of stars.

The following year the Swiss Fritz Zwicky (who also spent most of his career in the USA) measured the dynamics of the Coma cluster of galaxies, and found that the observed mass accounted for only about a tenth of what had to be there to provide the gravitational pull. He concluded that there has to be much more material there – in some dark form that does not show up in telescopes – or else the Coma cluster would simply disperse.

In the 1970s astronomers started to get odd results when they measured the rate at which stars orbited the centres of galaxies. They found that the stars far from the centre moved just as fast as those closer in. This is a very strange result indeed. Recall that in our solar system, the inner planets orbit faster than the outer ones. Mercury is tearing through space at about 30 miles (48 kilometres) a second in its orbit around the Sun, whereas Pluto can manage only a sluggish 3 miles (5 kilometres) a

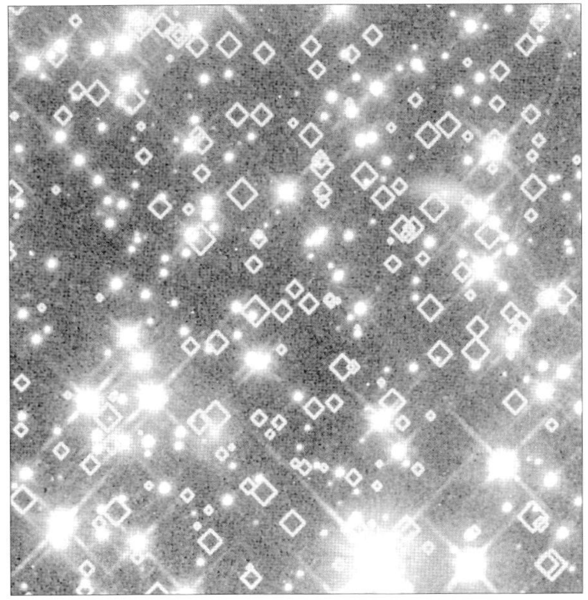

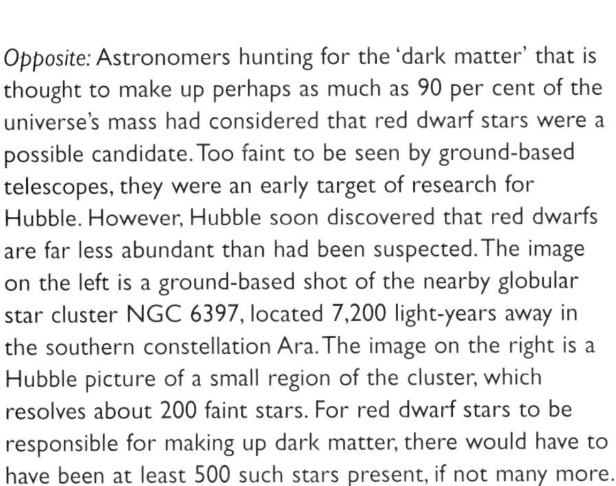

Opposite: Astronomers hunting for the 'dark matter' that is thought to make up perhaps as much as 90 per cent of the universe's mass had considered that red dwarf stars were a possible candidate. Too faint to be seen by ground-based telescopes, they were an early target of research for Hubble. However, Hubble soon discovered that red dwarfs are far less abundant than had been suspected. The image on the left is a ground-based shot of the nearby globular star cluster NGC 6397, located 7,200 light-years away in the southern constellation Ara. The image on the right is a Hubble picture of a small region of the cluster, which resolves about 200 faint stars. For red dwarf stars to be responsible for making up dark matter, there would have to have been at least 500 such stars present, if not many more.

Credit: F. Paresce (STScI and ESA), and NASA

Above: Scientists hunting for dark matter (*see caption, left*) analysed the images Hubble brought back from globular cluster, and simulated what they would have expected to see had faint stars been responsible for the matter which cannot be seen by ground-based telescopes, yet whose gravitational influence suggests it accounts for most of the matter in the universe. The diamonds added to the left image show the extent to which faint stars should have shown up; the true image on the right shows the degree to which the real faint stars are present. Indeed, there are so few faint stars that Hubble can see right through the formation, picking out distant galaxies located far behind it.

Credit: F. Paresce (STScI and ESA), and NASA

second – one tenth as fast. This is the sort of pattern that would be expected in any system of bodies orbiting a central mass; if it works for the solar system, it ought to work for the stars orbiting the centre of a galaxy. But it does not.

A more sophisticated approach is to consider the stability of the spiral arms of a galaxy like our own Milky Way. Calculations show that the spirals would break up if they contained nothing but the material that astronomers can observe in their telescopes.

The only conclusion that can be drawn from these observations is that there must be something – and something pretty massive – whose gravitational pull is interfering with the simple motion that one expects.

This mysterious stuff lurking in space has come to be known as 'dark matter'. Our entire galaxy, and the others whose rotation can be measured, simply must be embedded in a 'halo' of this dark matter. And, bizarrely, there may be much more dark matter in the universe than the ordinary stuff;

indeed, estimates range between five and ten times as much. If the latter is the case, then 90 per cent of the universe is made of material which astronomers cannot see through their telescopes. And this form of matter is dark not just to optical telescopes but all the way across the spectrum from radio waves to gamma rays.

The range of candidates put up for 'dark matter' is a testimony to the imagination of astronomers, cosmologists and physicists. They fall roughly into three categories: 'machos', 'wimps' and black holes.

Some more conservative thinkers suggest that the massive haloes round galaxies might just be red dwarf and brown dwarf stars. These are the machos: 'massive compact halo objects'. Red dwarfs, the theory goes, being dull stars, might shine so dimly that ground-based telescopes (if they picked them up at all) would confuse them with distant galaxies. Among observable stars, smaller ones tend to be more plentiful than really big ones in the galaxies, so maybe (so the speculation went) this trend continues below the limit of observation by ground-based telescopes. But as long ago as 1994 the Hubble Space Telescope ruled out red dwarfs as the main source of dark matter. It was capable of picking out red dwarfs that were more than 100 times dimmer than could be seen using ground-based telescopes, resolving them into sharp pinpoints characteristic of stars rather than confusing them with the fuzzy extended objects that are distant galaxies. But the results were disappointing to proponents of the red dwarf idea; so few were seen that red dwarfs can account for only 6 per cent of the mass of the halo of the Milky Way.

However, this does not rule out the possibility that the machos might be brown dwarfs, the wannabe stars that were just too small to catch alight with thermonuclear fire, and so do not shine out in the night sky. The only way to detect

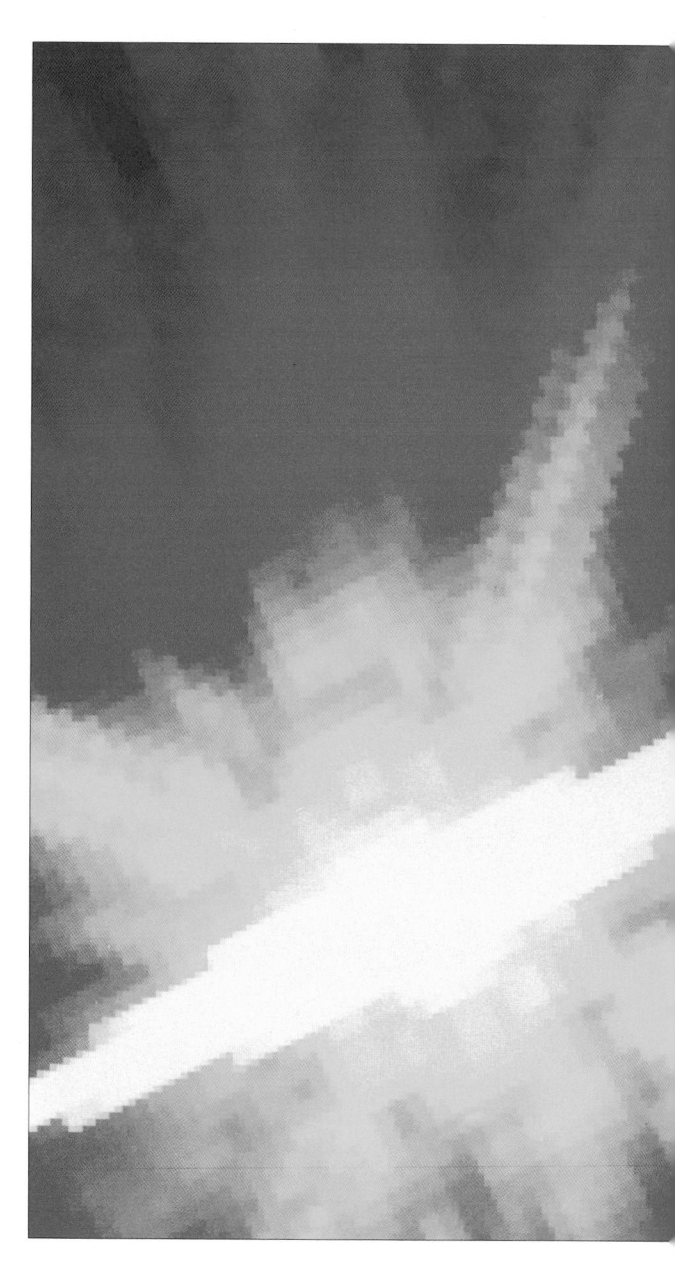

these objects is by a trick of their gravitation (yet another effect successfully predicted by Einstein's General Theory of Relativity), which can literally bend rays of light so that, in certain positions, they can act like a lens to focus the light from distant stars. An international team of astronomers has now found more than 100 such machos in our galaxy and in the Large Magellanic Cloud, after monitoring more than 1 million stars for any

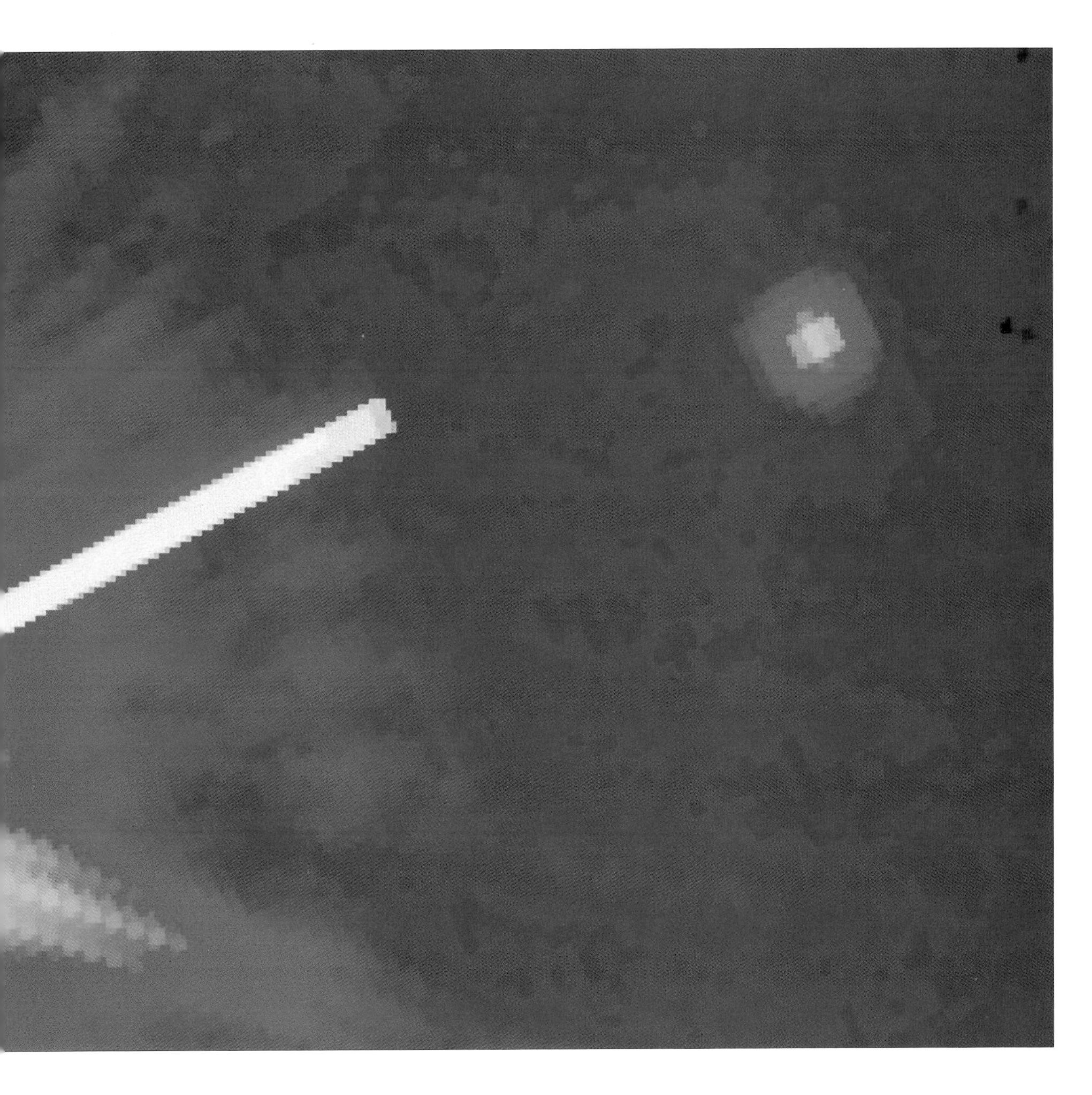

A red dwarf star, one of the smallest, least massive, and coolest stars ever seen. It is visible on the right of the image, and is called Gliese 105C. It is part of a binary system, and its companion Gliese 105A is on the left of the picture. The pair of stars are located 27 light-years away in the constellation Cetus. The Hubble image suggests that Gliese 105C is 25,000 times less bright that Gliese 105A – which helps explain why ground-based telescopes have had such difficulty in picking out these faint stars, the smallest celestial objects to actually qualify as true stars, shining by virtue of the nuclear reaction taking place in their interior. Were this red dwarf to be placed in our solar system, at the same distance from us as our Sun, it would shine only four times more brightly than the full Moon. It is estimated that Gliese 105C has about 8 to 9 per cent of the mass of our Sun, putting it pretty close to the theoretical lower limit at which there is sufficient mass to trigger the nuclear reaction that 'fires up' a star. Red dwarf stars, because they are cooler and less energetic, are extremely economical with their fuel; they may end up shining for 100 billion years.

Credit: D. Golimowski (Johns Hopkins University), and NASA

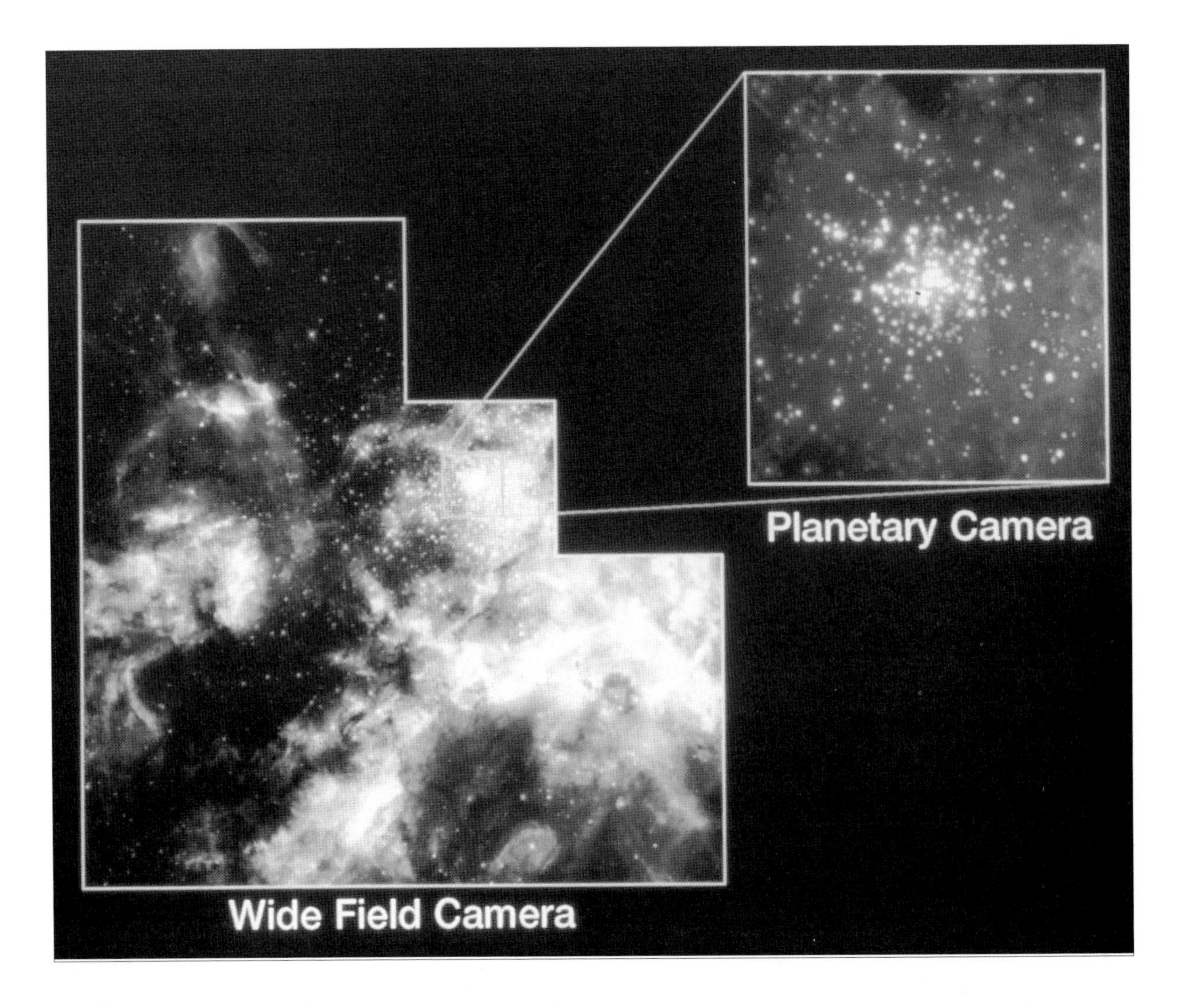

One of the Hubble Space Telescope's primary missions is to give astronomers the evidence they need to confirm or deny theoretical models about how stars and galaxies are born. Once this process is properly understood, scientists will have a much better chance of predicting how the universe will evolve. The main picture shows a star-forming region in the 30 Doradus nebula, part of a small galaxy close to our own Milky Way called the Large Magellanic Cloud, which is located 160,000 light-years from Earth. The cloud of dust and gas in the picture is made up mainly of ionised hydrogen, which is glowing because it has been energised and excited by ultraviolet light coming from hundreds of young, bright and massive stars clustered together. The close-up taken by Hubble of the star cluster, known as R136, showed that it contains more than 3,000 stars which have formed recently from out of the cloud.

Credit: NASA

change in the brightness that might result from this gravitational lensing.

Other theorists, believing that Big Bang put a limit to the amount of conventional matter that can be present in the universe, have plumped for wimps – 'weakly interacting massive particles' – as the candidates for dark matter. These are exotic subnuclear particles, only a few of which have ever been observed in the laboratory. One possible candidate is the neutrino, a tiny electrically neutral particle which is produced in some types of radioactive decay. This candidate has the merit that it actually does exist in reality and is not the figment of a theorist's overactive mind. But there is a snag: the

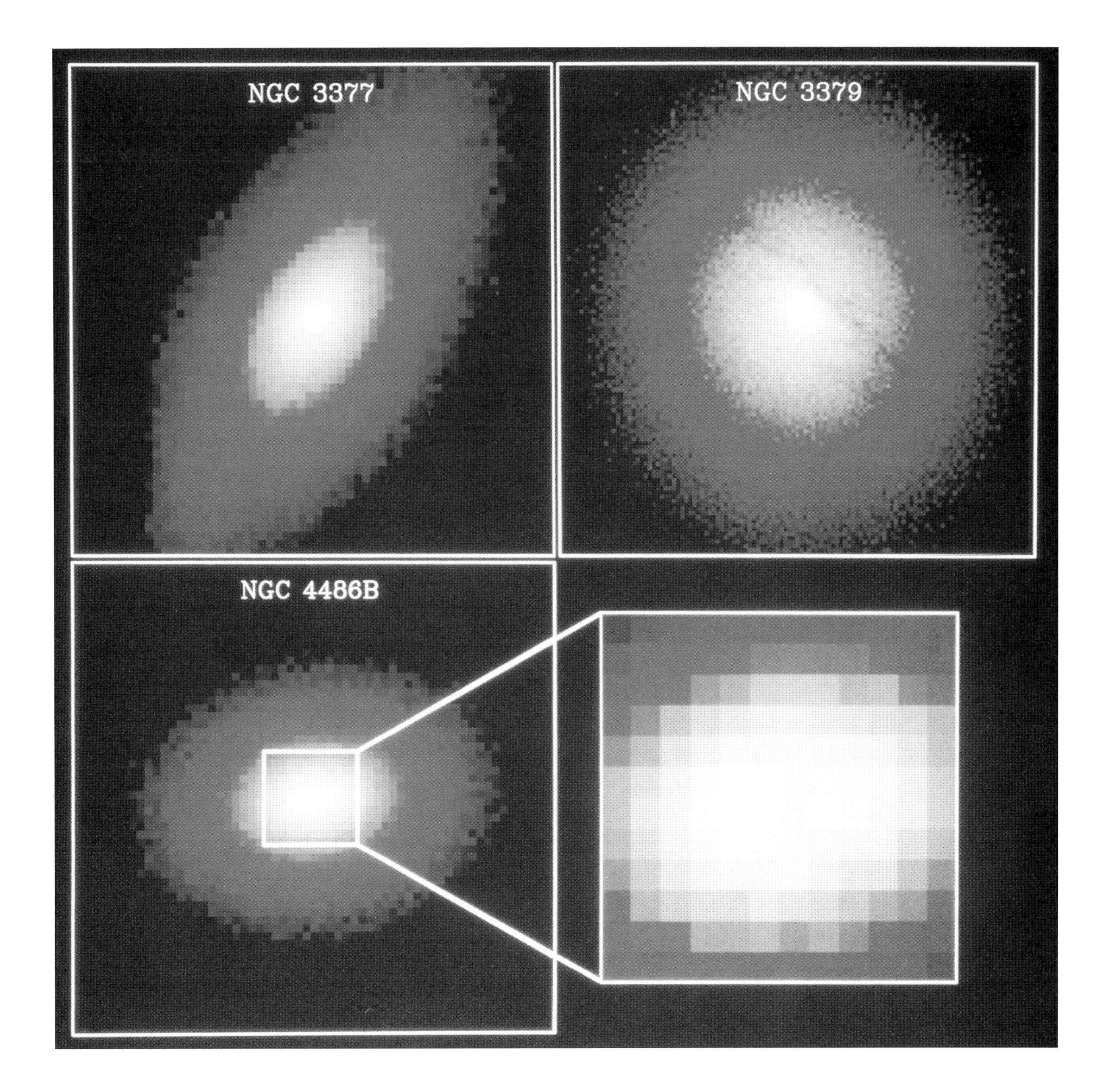

Black holes have, until recently, been the subject of theoretical exposition (they were predicted in Einstein's Theory of General Relativity, but were first mooted as long ago as the eighteenth Century) and extremely imaginative (not to say wild) speculation. Almost all possible roles have been assigned to them: entrances to 'wormholes', cosmic superhighways that would allow faster-than-light journeys; the key to time travel; entrances to parallel universes; and so on. However, Hubble has come up with more and more hard evidence not only that black holes exist, but that they may be very common – so common, in fact, that most galaxies (including our own Milky Way) may have one or more black holes lurking at their cores. The image above shows three galaxies (NGC 3377, NGC 3379, and NGC 4486B) where Hubble has detected the presence of central, supermassive black holes. The one at the heart of NGC 4486B, for example, is thought to have 500 million times as much mass as our Sun. In some galaxies, black holes are thought to be dormant, and relatively quiescent. In others however, they are feeding voraciously, gobbling up entire stars one after another.

Credit: Karl Gebhardt (University of Michigan) and Tod Lauer (NOAO), and NASA

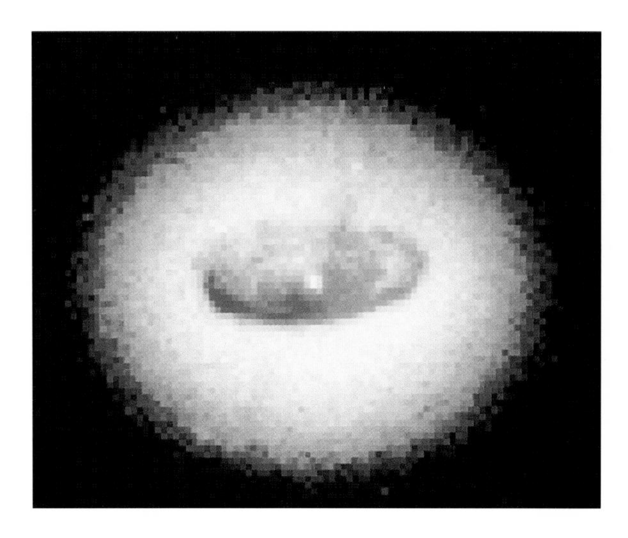

Galaxy NGC 4261, where Hubble has identified a massive black hole. The image clearly shows a huge spiral of gas and dust, 800 light-years across, which is fuelling the black hole. The galaxy is located about 100 million light-years away from Earth.

Credit: H. Ford and L. Ferrarese (Johns Hopkins University), and NASA

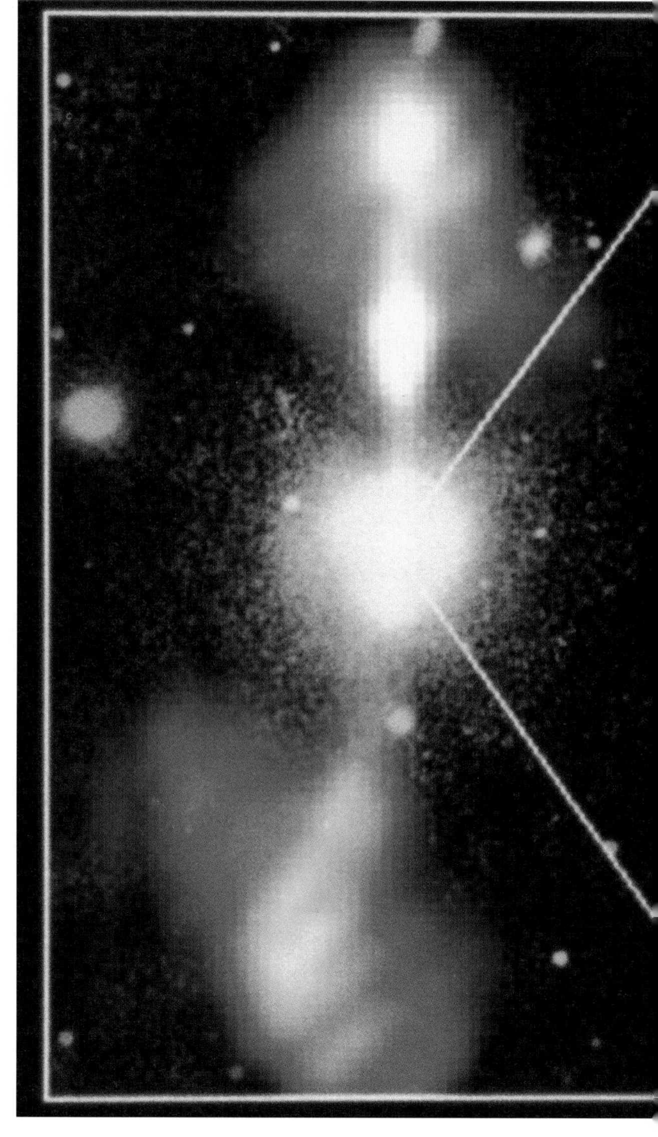

neutrino is traditionally regarded as being without mass, like the 'particle' of electro-magnetic radiation, the photon. If neutrinos have no mass then it stands to reason that they cannot account for the missing mass of the universe. This has led some theorists to suggest that they might have a tiny mass, too small to measure in the laboratory, but which mounts up to the universe's missing mass simply because there are so many wimps permeating the entire universe.

The final speculation is that the entire universe is permeated not with exotic subnuclear particles but with primordial black holes. This idea goes to the heart of the uncertainty over galaxy formation. Perhaps, the theory goes, some of the initial clouds of gas in the universe after Big Bang were so immense that they did not fragment into lumps that later became stars, which then collected into galaxies. Perhaps many clouds collapsed directly into supermassive stars which rapidly evolved into black holes, even before galaxies were formed. These could have been as massive as 1 million Suns and could have formed within a few million years of the universe's creation. The gas that remained would fall under the gravitational influence of these black holes and as the stars formed out of this gas they would naturally evolve into galaxies embedded in dark haloes of black holes. And if one such black hole were to find its way to the nucleus of a particular galaxy, the ingredients for a quasar (so much more numerous

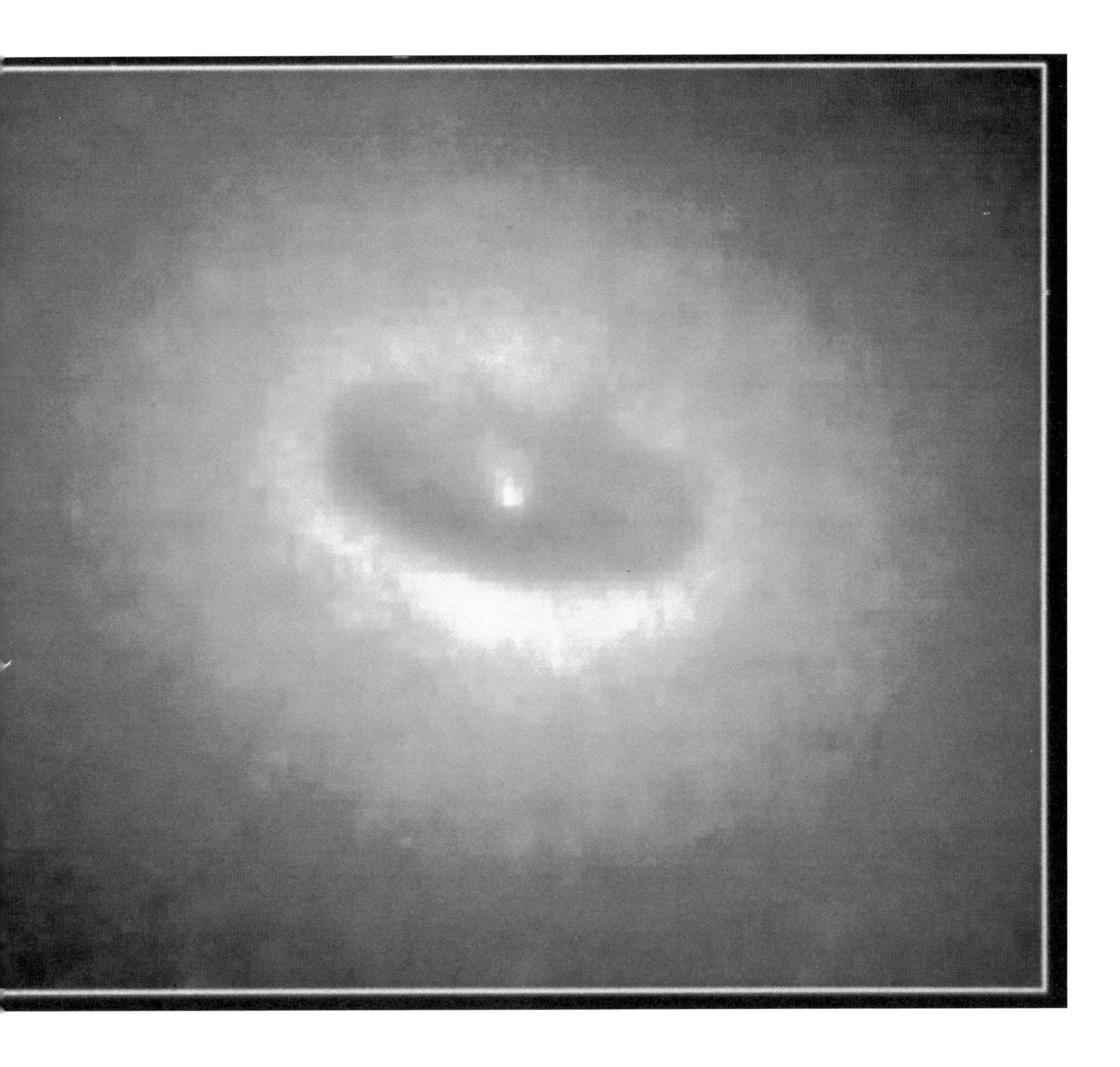

An earlier Hubble view of the core of galaxy NGC 4261, which is placed in context by being shown next to a ground-based image (above, left) of the entire galaxy. Studies of the disk of gas and dust imaged by Hubble have used the speed at which the material is moving around the black hole to produce an estimate of the size of the unseen black hole. This is now calculated to have a mass equal to 1.2 billion times that of our Sun – yet the entire black hole occupies a space not much larger than our solar system. The disk surrounding the black hole is estimated to have material with enough mass to create 100,000 stars like our Sun. Analysis of later Hubble images of this galaxy have shown that the black hole and its attendant disk are actually offset from the centre of the galaxy, suggesting that some sort of still-unknown dynamic interaction is taking place. We still do not know the exact relationship between black holes and galaxies. One theory has it that after galaxies formed, concentrations of stars at their cores became so dense that they collapsed to form a black hole. Others speculate that the collapse of a massive star into a black hole draws all the material in the neighbourhood towards it, creating a mass of dust and gas which in turn coalesces to form stars, thus eventually creating a galaxy. It is hoped that further studies by Hubble – and generations of new space-borne instruments to come – will produce further evidence that will clear up this mystery.
Credit: H. Ford and L. Ferrarese (Johns Hopkins University), and NASA

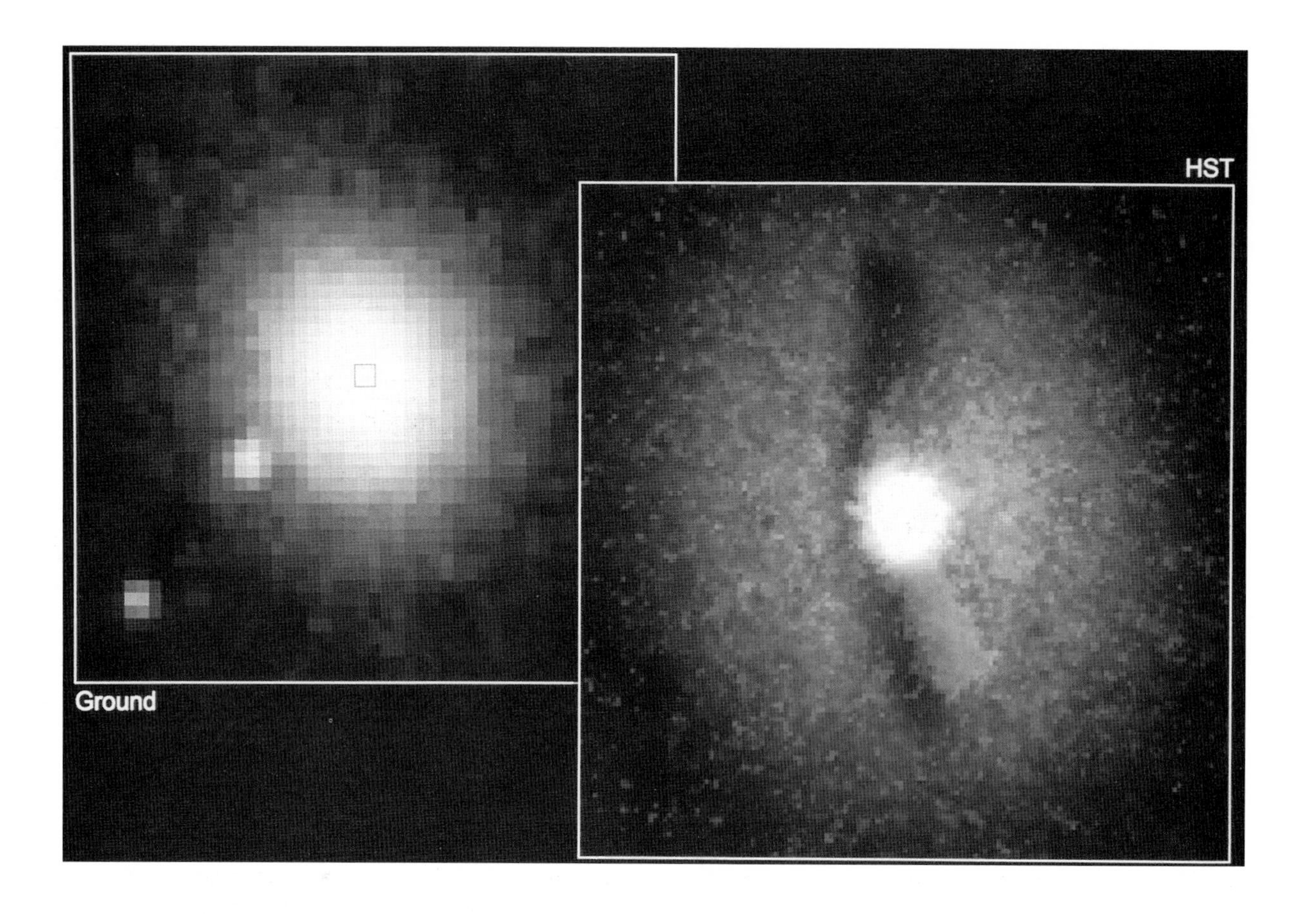

Hubble provides a sight never seen before – a warped disk of material circling a black hole in galaxy NGC 6251, and lit up by a flood of ultraviolet light generated by the hot gas trapped around the black hole. The dark shadow is the disk of dust seen in visible light, whereas the bright streaks are the reflections of the ultraviolet light. Because only one side of the disk is illuminated, astronomers conclude the disk must be warped, like the brim of a hat.

Credit: P. Crane and J. Vernet (European Southern Observatory), and NASA

in the early universe than today) would be in place.

According to Einstein's General Theory of Relativity, gravity warps the very shape of space and time, dragging it into a curve. A black hole is the site of perhaps the most extreme warping of all; eventually space and time close in on themselves. Anything falling into a black hole falls to a point of singularity at the centre, the point at which space, time, and matter all cease to exist. But ingenious physicists have come up with ways of solving Einstein's equations in which it might be possible to escape such an unhappy fate. There is the possibility that one could pass through a black hole (avoiding the singularity) and pass through the other side, emerging out of a 'white hole' in another part of the universe (or possibly even in a different universe). A white hole is the inverse of a black hole: spewing out matter and energy, rather than sucking it in, the white hole would be rather like a miniature version of the Big Bang. Three things have to be said. The first is that there is no evidence (yet) for the existence of white holes. The second is that even in this speculation-ridden field, most sober cosmologists are sceptical about the possibility that anything falling into a black hole could escape being squashed into oblivion at the point of singularity. And the third is that the

Scientists probing the furthermost recesses of the universe have been able to use 'gravitational lenses' to see further than ever before. This image shows the galaxy cluster Abell 2218, where the concentration of mass is so great that light rays coming through it are bent and concentrated, as if they were passing through a huge lens. The thin arc-like streaks are images of distant galaxies, lying far beyond Abell 2218.

Credits: W. Couch (University of New South Wales), R. Ellis (Cambridge University), and NASA

gravitational pull increases so sharply close to a black hole that tidal forces acting just across the distance between a traveller's head and her feet might well tear her apart even before she got anywhere close to the point of singularity.

Nothing daunted, the late American astronomer Carl Sagan suggest that black holes might be used as a way of 'commuting' between different parts of the universe.

The distances between the stars are immense; the distances between the galaxies are truly vast, beyond the capacity of the human mind to comprehend. No conventional technology could get us even to the nearest star and back again within a human lifetime. Suppose for a moment we could adopt a *Star Trek* teleporting system (assuming any such thing were possible) whereby our atoms were disassembled and their specification transmitted across space at the speed of light to be reassembled at the other end. The fastest that this information could travel is at the speed of light. This still means that a round trip to Proxima Centauri – the star nearest to the Sun – would take more than eight years. One of the bizarre effects of Einstein's Special Theory of Relativity is that for the person actually undertaking a journey at the speed of light, time would not appear to pass at all and the journey would to them appear to have taken quite literally next to no time. A round trip to view the proplyds – the formation of planetary systems – in the 'nearby' Orion nebula would take 3,000 years, a period virtually as long as the whole of recorded human history. Travelling to the Andromeda galaxy and back would occupy a period longer than that since the evolution of recognisable humans on the face of the Earth. Who would set out on such a journey knowing that they would return to a civilisation that would have changed by more than 4 million years of development? Or which might have passed out of existence altogether through ennui or self-destruction?

So there is an undoubted attraction to Sagan's suggestion that the link between a black hole at one end and a white hole at the other might offer a 'wormhole' through space that would allow future travellers to move from one region to another faster than any other medium. Even if such wormholes opened up, calculations in General Relativity suggest that they would pinch off – the hole would slam shut – as soon as anyone entered it. But the marriage of quantum physics with Einstein's relativity theory suggests that it might be possible to keep the hole open. Could this be the 'warp drive' also beloved of *Star Trek*? And if they can be used to travel across vast expanses of space, the theorists argue, could wormholes not also be used to travel in time?

Certainly, there are theoretical solutions, discovered as long ago as the 1930s, of relativity which suggest that time travel might be possible. But there are two problems here. One is that there is a great difference between sending a single particle – a proton, say – back in time, and sending a human being. A human being is a highly ordered composition of matter and it is rather important to us that we remain that way – simply piling up the

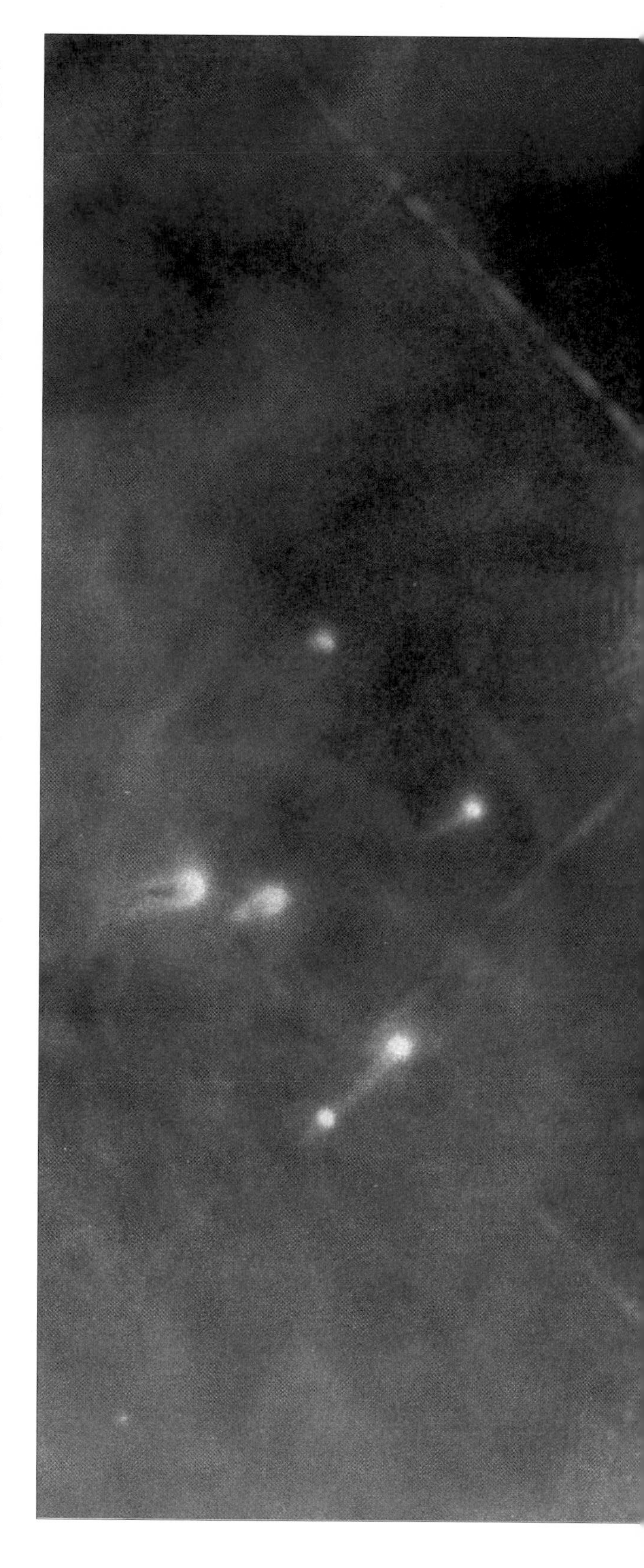

Studies of the Trapezium cluster of stars in the Orion nebula suggest that many protoplanetary disks surrounding young stars may have a tough time lasting long enough to form planets. Radiation from the bright, massive nearby stars in this picture is eroding the outer layers of the protoplanetary disks (visible as blobs in the picture), causing them to evaporate off the main disks in the form of ionised gas, which is heated to 10,000 degrees Celsius and glows as a result.

Credits: D. Johnstone (Canadian Institute for Theoretical Astrophysics, University of Toronto), and NASA

chemical ingredients for a human being does not make a man or a woman. It may be that all the 'information' that constitutes a most important physical part of being human would be lost in transit through a wormhole, and that would be an utter and terminal disaster for the would-be traveller.

The second objection is one of principle. Time travel would interfere with our notions of cause and effect. This is best expressed in the old paradox: what would happen if you went back in time and inadvertently killed your grandparents before your parents were born? If your parents were never born, you could not have been born and so you could not have gone back in time inadvertently to kill your grandparents, but somehow you did . . . Or you may decide to travel back in time to kill some infamous dictator before he could wreak his evil – but you *must* have failed, because had you successfully killed him he would never have posed the threat that would have prompted you to travel back in time to kill him. . . It may be, as the mathematical physicist Stephen Hawking has suggested, that there is a cosmic censorship principle operating to prevent paradoxes such as these.

Setting aside the wild speculations about time travel, even getting off Earth and out of the solar system presents scientists with serious problems, especially when it comes to the question: 'Are we alone?' A better question might be: 'If we are not alone, will we ever meet?' The impracticability of travelling very far from our own planet in a reasonable period compared to the average human lifetime has sobering implications for the possibility that we might ever encounter other forms of life and other civilisations. The enormous numbers that are needed to describe the universe simultaneously make it likely that there are indeed other forms of life than our own out there; they also make it very likely that we will never have physical contact with them.

In the first chapter of this book the rough estimate was given that there are around 100 billion stars in our galaxy. No one can count how many galaxies there might be in the universe, but it could be anything from 50 billion to 200 billion. This would mean that in total there are more than 20,000,000,000,000,000,000,000 stars – 20,000 billion billion – in the universe. Evidence from our own galaxy suggests that many stars have planets; but suppose only one star in a million were to have planets, and only on one planetary system in a million did life evolve, and in only one out of a million cases did the evolution of life progress to the development of intelligence and technological civilisation. That calculation of ever-dwindling possibilities still means that, today, there would be 20,000 planets across the universe teeming with intelligent life.

As was mentioned at the beginning of the last chapter, most of the stars in the universe are getting old, whereas our Sun is a comparative latecomer; so it is possible that the majority of these imaginary 20,000 planets with intelligent life have further-developed civilisations than ourselves. But then the other aspect of the enormous numbers at work in our universe comes into play to mock any hope that we might ever know of these other citadels of intelligent life, let alone travel to them.

Consider the nearest candidate for such life, Barnard's Star, which may have planets orbiting it. (Even so, as a red dwarf star only one-hundredth as bright as the Sun, its planets would be very cold and gloomy places; any life there would not be as we know it.) Barnard's Star lies six light-years away. At the speeds attained by the Voyager and Pioneer space probes on their way out of the solar system (around 25,000 miles/40,000 kilometres an hour) it would take 136,000 years to get to Barnard's Star. Even were it possible to design a spaceship that could travel at one-tenth the speed of light (i.e. at 18,600 miles/30,000 kilometres per second) the

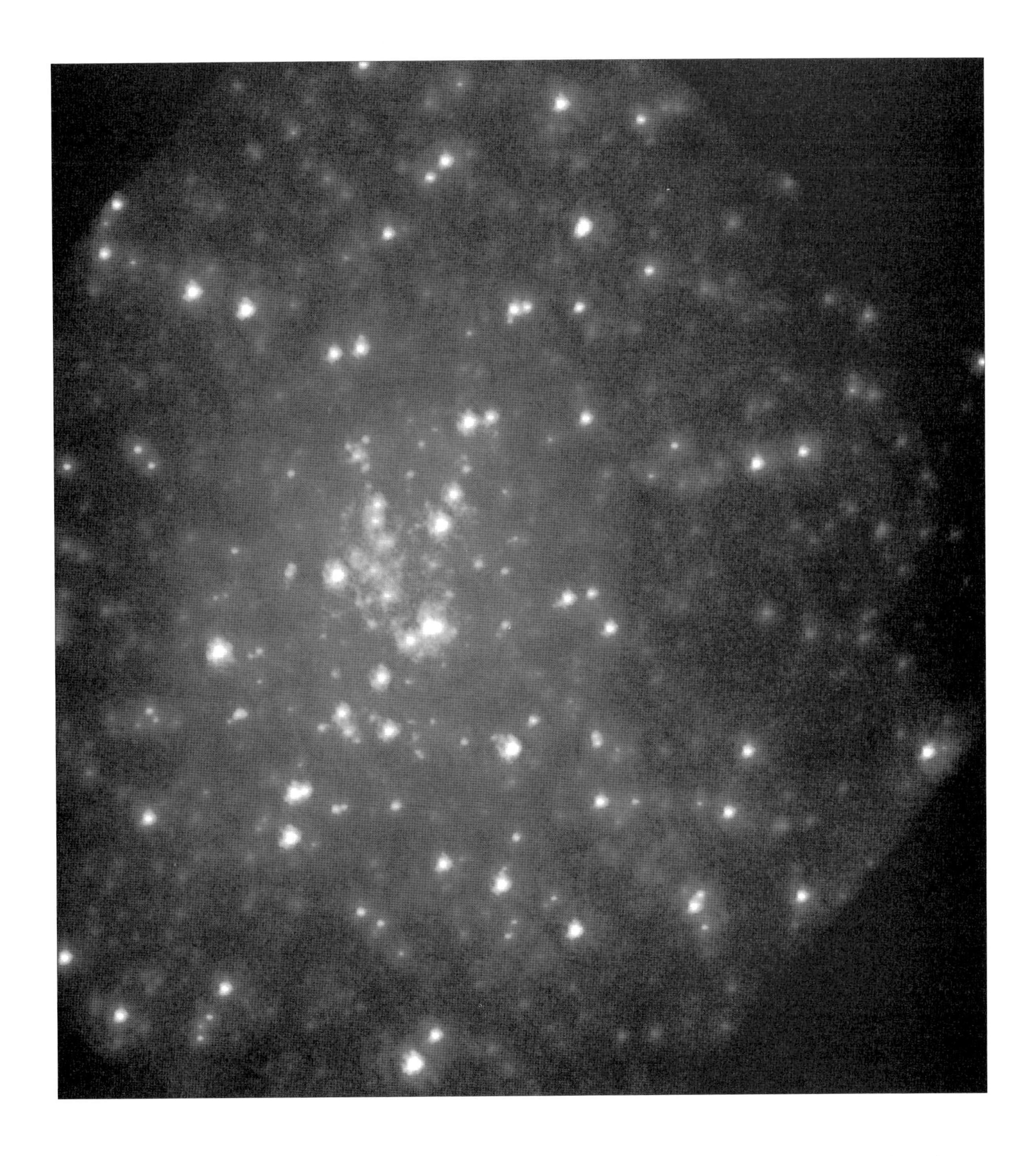

Hubble found a population of fifteen of intensely hot blue stars at the centre of M15, one of the most dense globular clusters of stars seen to date. It appears that the stars are so closely crowded together that their gravity has stripped away the outer layers of gas that normally surround a star's core. Astronomers believe that these blue stars are the remaining exposed cores of red giants which lost their outer layers to passing neighbours. M15 is visible to the naked eye as a hazy smear about one-third of the Moon's diameter, located 30,000 light-years away in the constellation Pegasus.

Credits: Guido De Marchi and Francesco Paresce (STScI and ESA), and NASA

outward journey alone would take sixty years. There is no hope of being able to conduct a physical search even of our nearest neighbour stars in the attempt to find extra-terrestrial life.

Such considerations have led some astronomers to conclude that the only way in which to search for extraterrestrial life is by employing the world's radio telescopes to listen for signals that might betray the presence of a technological civilisation. Not all these signals need necessarily be attempts at communication with other worlds, some of them could be the domestic radio transmissions that have also radiated away into interstellar space. For example, if there are civilisations out there about forty light-years from Earth, they should just be picking up the first episodes of the *I Love Lucy* show. Quite what they will make of such signals, mixed up perhaps with the broadcast of the coronation of Queen

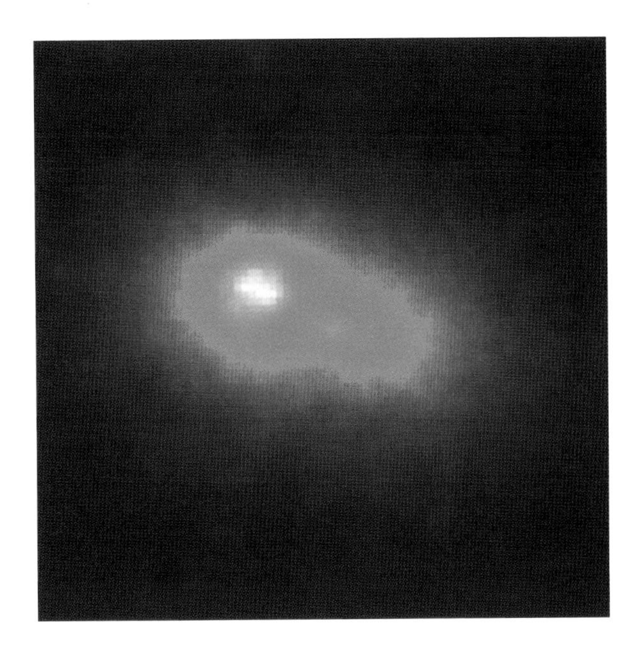

Our own Milky Way galaxy must look like this image of Andromeda, a characteristic spiral galaxy (*left*, taken from a ground-based telescope). Even though this Hubble photograph (*above*) of the nucleus of the galaxy was taken before the repair mission which sharpened the telescope's vision, the image shows how closely the space telescope can peer into the core of our closest neighbour.

Credit: T. R. Lauer (NOAO), NASA

Elizabeth II in Britain (one of the first ever live outside broadcasts on television), is something that will surely tax their intelligence.

This is a search that falls to astronomical equipment other than the Hubble Space Telescope. As director of the SETI Institute (Search for Extra-Terrestrial Intelligence), Jill Tarter scans the skies listening for signals that indicate life might exist on other planets. Currently, she is using time on the world's largest radio telescope, at the Arecibo Observatory in Puerto Rico. At 1,000 feet (305 metres) across, Arecibo's dish is about sixteen times the size of Britain's Jodrell Bank radio telescope. The large area allows radio-astronomers to collect very weak radio signals and concentrate them. Arecibo is so powerful that it could listen in on a mobile phone conversation on the planet Venus, or detect an object the size of a golf ball on the Moon.

While the whole of the Jodrell Bank telescope can be steered to point at any object in the sky, Arecibo is fixed solid in a natural bowl in the hills of north-west Puerto Rico. Carl Sagan and Frank Drake did the first search for extraterrestrial life at Arecibo in 1975–6, but found no signs of life. In 1993 NASA stopped funding the SETI project because of political pressure from conservative politicians who thought that looking for life on other planets was a waste of public money. Now, the SETI Institute runs on private money. But Jill Tarter wants a larger ear than Arecibo; she has asked for the Moon. Dr Tarter has predicted that eventually it will be necessary to put radio telescopes on the far side of the Moon, away from Earth, because humans are polluting our radio spectrum at all wavelengths with ever-increasing numbers of signals emitted by instruments from telecommunications satellites to cellular phones.

Perhaps that is an appropriate place at which to close this look at the universe through the orbiting eye of the Hubble Space Telescope. It has settled some outstanding questions that have troubled astronomers for decades; it is close to finding a definitive value for the Hubble Constant, one of the most important numbers in the whole of science. But it is also opening up new questions, some of which the Hubble Space Telescope itself will be unable to answer alone.

For centuries astronomers looked only at the visible light from the stars. The privilege of the scientists of the post-war generation has been that they could view the stars at all parts of the spectrum: from radio waves, through infra-red and visible light, to ultraviolet, X-rays and gamma-rays. The Hubble Space Telescope can see a little bit into both the infra-red and the ultraviolet, and this

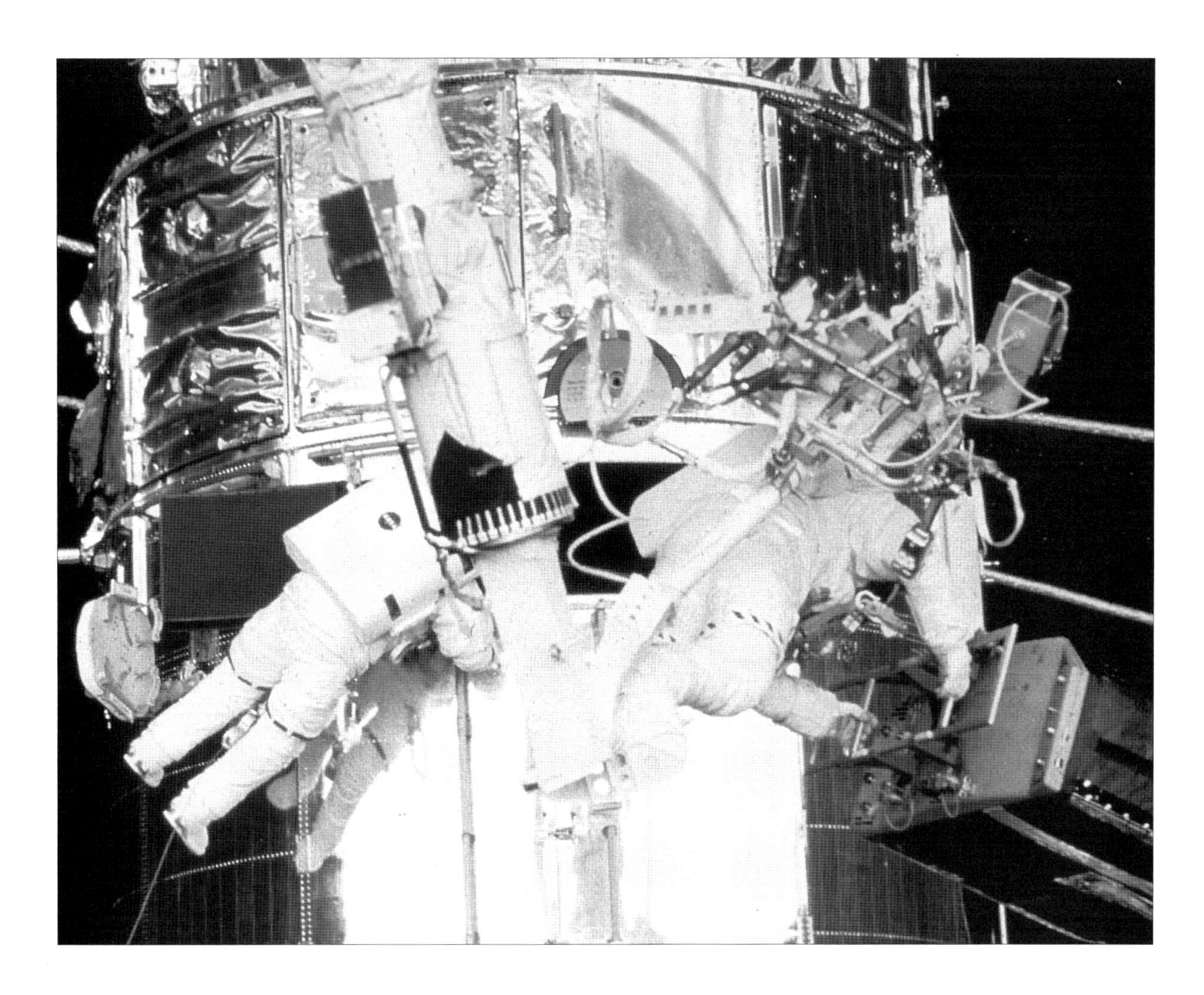

Above and left: Astronauts spent five days engaged in EVAs ('extra-vehicular activities' – i.e. spacewalks) during the second servicing mission to the Hubble Space Telescope in February 1997. This mission added two new instruments to Hubble's armoury; the near infra-red camera and multi-object spectrometer (NICMOS) and the space telescope imaging spectrograph (STIS). In addition, the astronauts replaced a number of components aboard Hubble, allowing it to continue its work in complete scientific health until the next servicing mission in 1999.

Credit: NASA

ability has proved crucial in providing some of the detailed numbers to answer the astronomers' questions. Scheduled servicing missions will further improve this capacity.

Eventually, however, even the Hubble Space Telescope will wear out and become redundant. Technically, it was built to last only until the year 2005. Although it now looks as if it might last longer, NASA is already laying out its plans for a replacement, the so-called Next Generation Telescope. This will have an even larger mirror (more than 13 feet/4 metres across) but should cost less than $1.5 billion. It will be designed particularly to pick up infra-red radiation and should be orbiting by the year 2007. Doubtless this too will solve some problems, and raise fresh ones. There is no limit to human curiosity, and to the questions that can be asked. But there is a limit to the satisfaction of this curiosity.

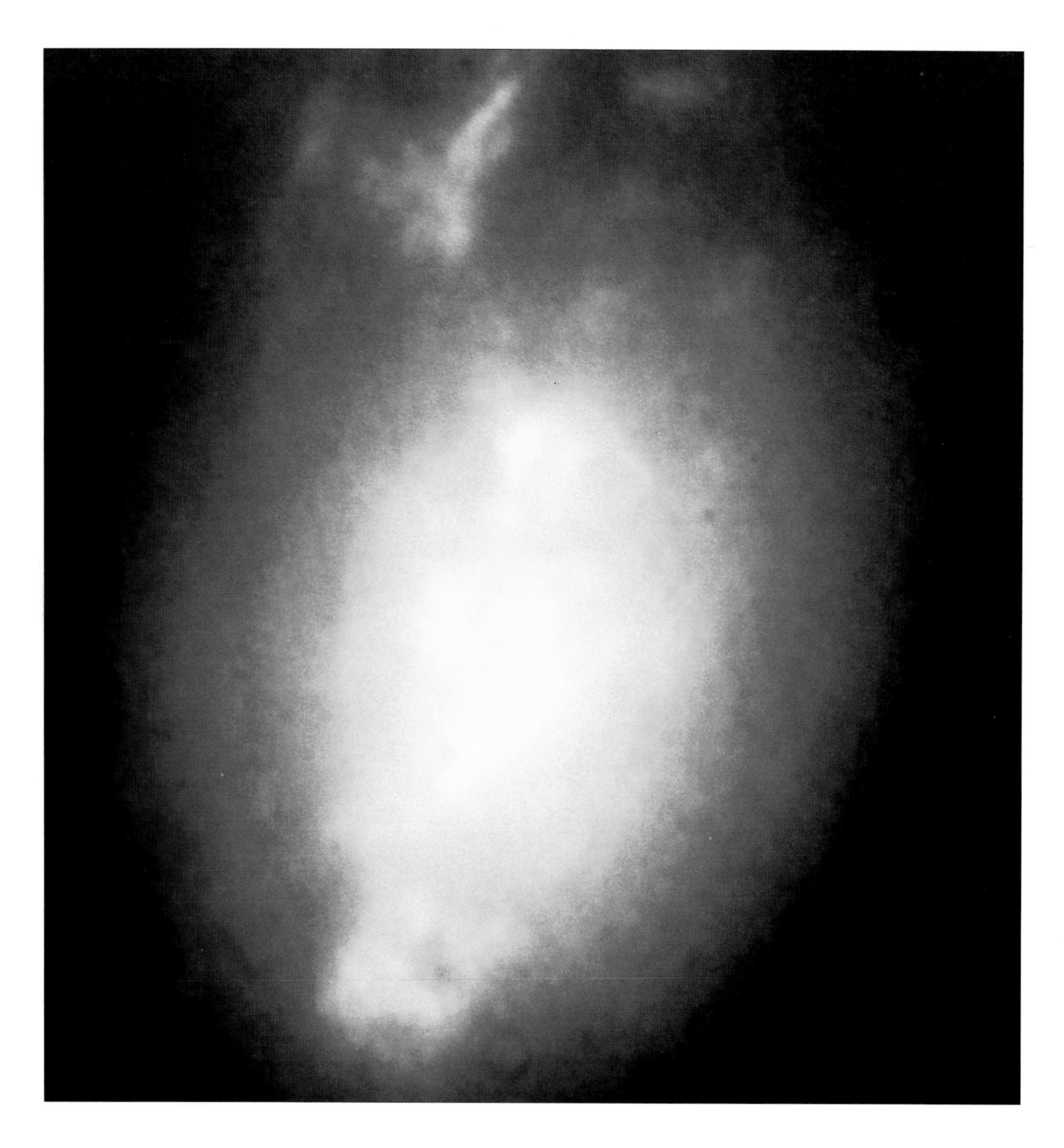

Above and opposite: The centre of the active Seyfert galaxy NGC 1068. About 60 million light years away it shines with the brightness of a billion Suns. Its core must contain a 'super massive' black-hole with a mass 100 million times that of the Sun.

Credit: D. Macchetto (ESA and (STScI) W. Sparks and A. Capetti (STScI).

One of the triumphs of science is that it has been so unreasonably successful. It is absurd to think that we, sitting here on the planet Earth, can work out the detailed mathematics of the processes going on in the interior of the stars. And yet we have. There is every reason to believe that the outstanding questions and mysteries we have touched

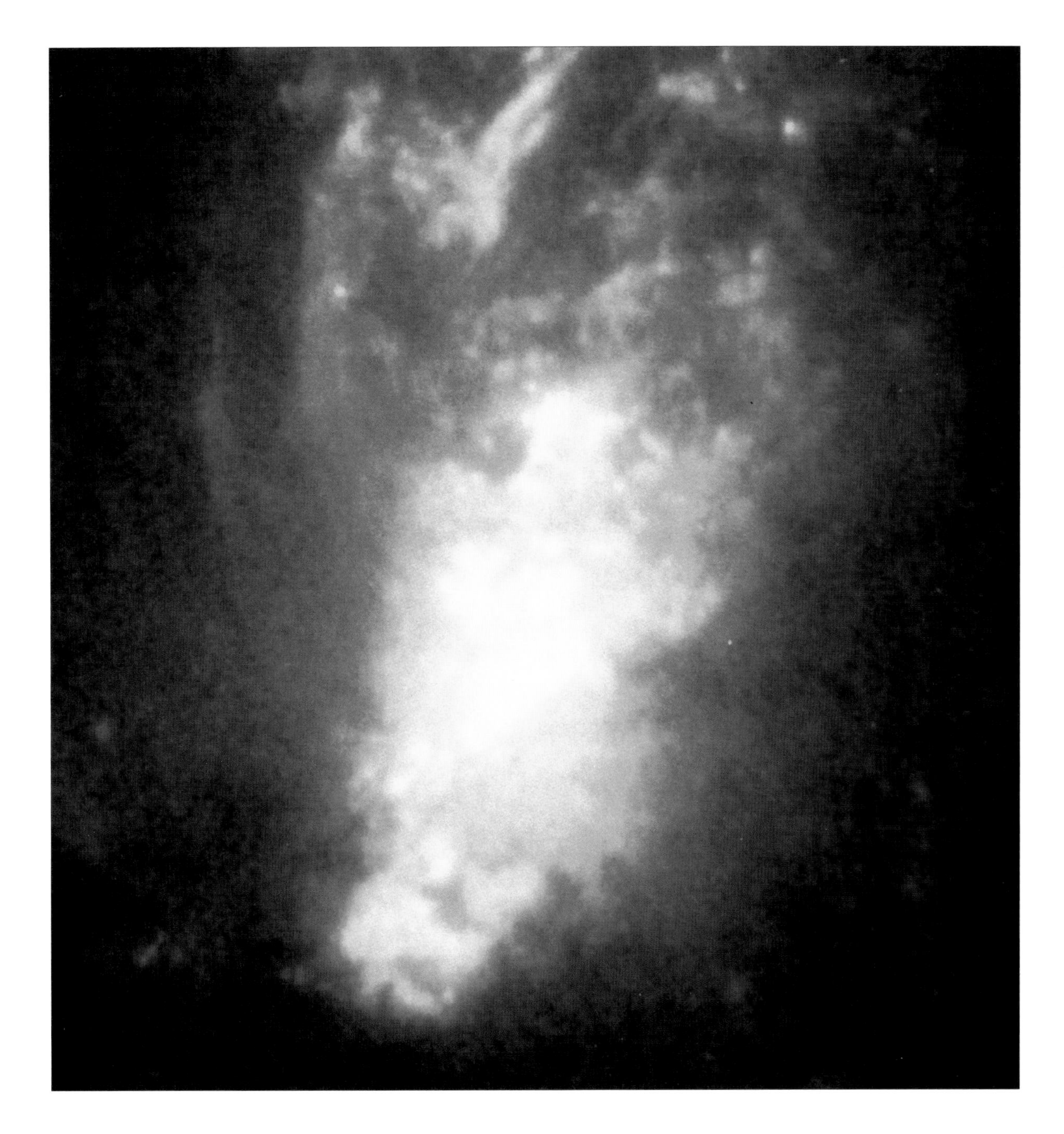

upon could, one day, also be answered. There is no barrier to the prowess of human imagination and human reasoning – except for one.

For the past decade or two, governments everywhere have shown themselves increasingly hostile to spending money on science and technology. The 'big questions' no longer interest them, nor does the idea that the human spirit can find expression and fulfilment through the details of observing the skies and working out mathematical models of how the very place in which we live came to be. In recent years the idea has grown up that the proper thing to spend money on is making yet more money. Curiosity about the natural world has been one of the deepest characteristics of our civilisation for centuries; but while it might tell us about our proper place in the scheme of things, it does not turn a financial profit.

The next generation of space telescopes will cost serious money and, in the end, this may prove to be the final horizon beyond which no astronomer can look. Not the edge of the observable universe, but the bottom line of the balanced budget.

This chapter has focused on some of the unanswered questions about the cosmos. It has entertained some speculations about the nature of dark matter, the possibility of wormholes in space and the existence of extraterrestrial space. These will remain speculations without detailed observations of the universe in which we live. Many of the explanations in the previous chapters are qualitative accounts which do not (and, in a book intended for a general readership, cannot) do justice to the lengthy and laborious mathematical calculations by which astronomers assure themselves that they have got their ideas right. It is only via this work – this long, arduous and expensive work – that we can state with any degree of certainty that our knowledge is real and true.

Edwin Hubble himself described his own discoveries in a celebrated book published in 1936 called *The Realm of the Nebulae*. He closed the book with an appeal for ever more, and ever more precise, observations of the heavens: 'With increasing distance our knowledge fades and fades rapidly. Eventually we reach the dim boundary, the utmost limits of our telescope. There we measure shadows, and we search among ghostly errors of measurement for landmarks that are scarcely more substantial. The search will continue. Not until the empirical resources are exhausted need we pass on to the dreamy realm of speculation.'

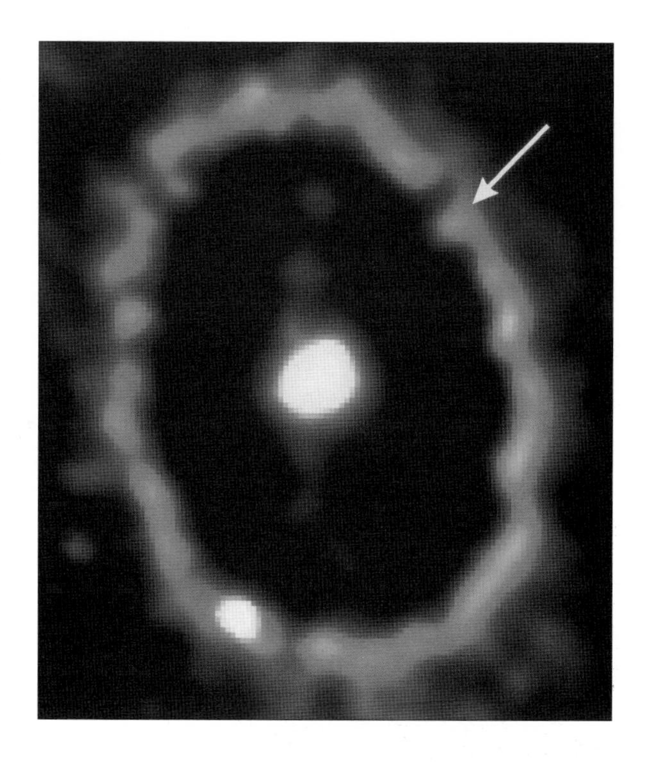

Above and right: Far from being unchanging or unimaginably slow, some cosmic events happen fast. Part of the ring of gas produced by supernova 1987A – the first naked-eye supernova for 400 years – has visibly brightened in just three years. This is the site of a collision between a blast wave and the innermost parts of the circumstellar ring, heating the gas.

Credit: P. Garnavich (Harvard-Smithsonian CfA) and NASA

Overleaf: Human courage and human technology alone in the immensity of the void. There is so much more for us to learn about the universe in which we live. But what is remarkable is that we already know so much.

Credit: NASA

Let us hope that Hubble's confidence was right, and that the search will continue. Without money to build new telescopes, humanity will end up unable to distinguish reality from the shadows and ghosts of error in its dreamy speculations.

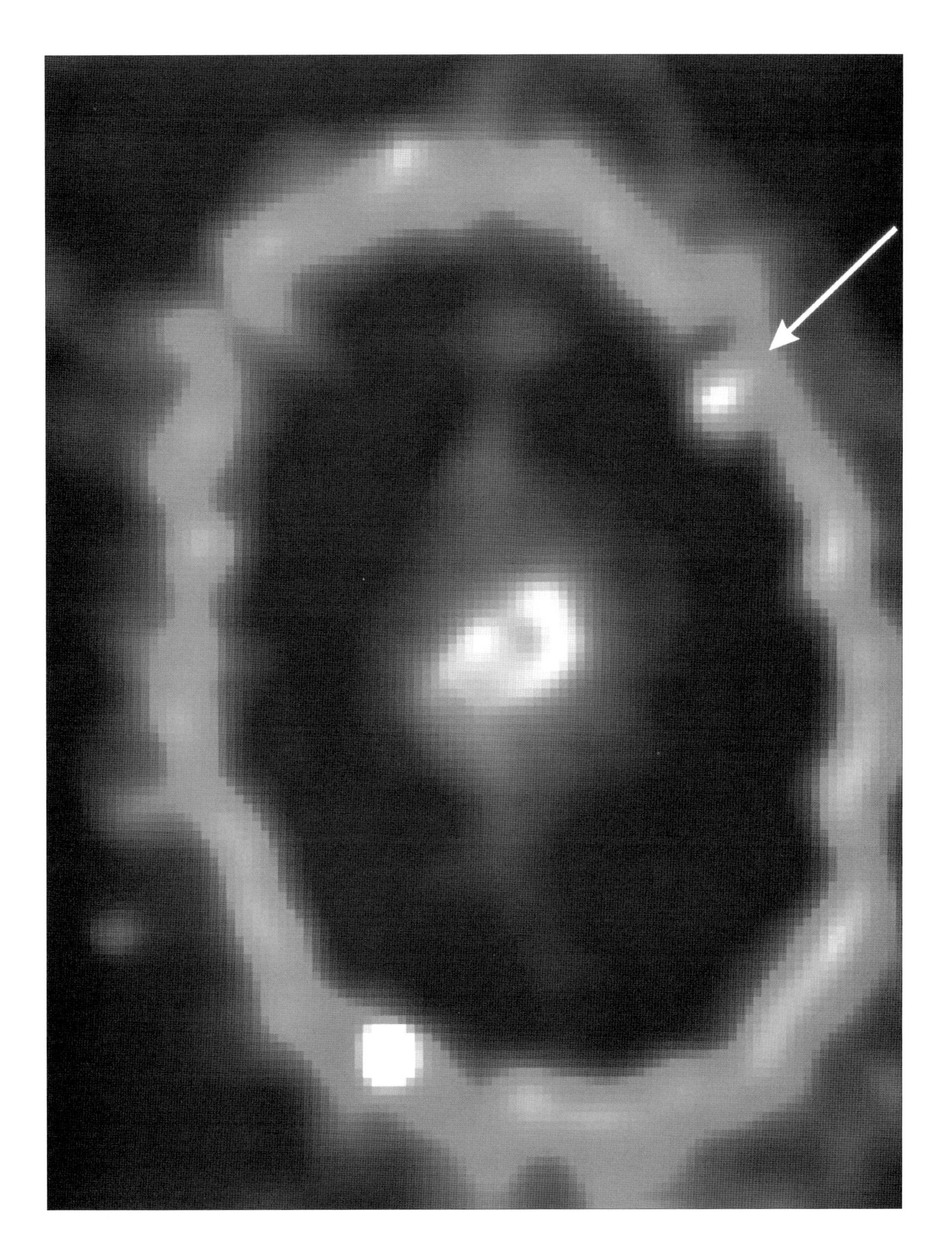